EINSTEIN'S UNFINISHED DREAM

DR. DON LINCOLN

EINSTEIN'S UNFINISHED DREAM

Practical Progress Towards a Theory of Everything

OXFORD
UNIVERSITY PRESS

OXFORD
UNIVERSITY PRESS

Oxford University Press is a department of the University of Oxford. It furthers the University's objective of excellence in research, scholarship, and education by publishing worldwide. Oxford is a registered trade mark of Oxford University Press in the UK and certain other countries.

Published in the United States of America by Oxford University Press
198 Madison Avenue, New York, NY 10016, United States of America.

Library of Congress Control Number: 2022058620
ISBN 978–0–19–763803–3

DOI: 10.1093/oso/9780197638033.001.0001

Printed by Integrated Books International, United States of America

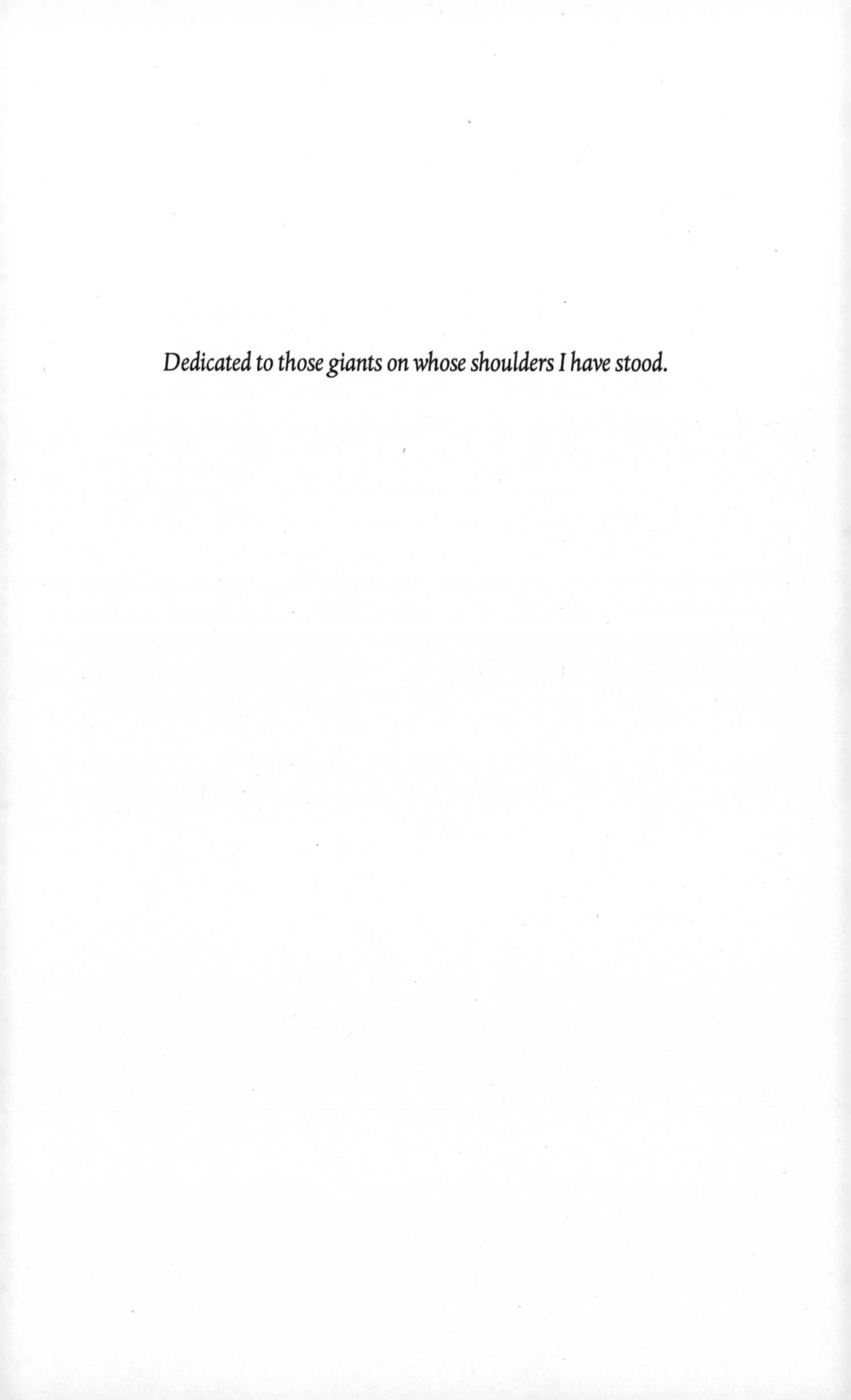

Dedicated to those giants on whose shoulders I have stood.

CONTENTS

FOREWORD

When I was just a teen, I was fascinated by what one might call "big questions," ancient questions of life and death and existence. These are topics that have historically been the province of theology and philosophy. However, over the centuries, the juggernaut that we call the scientific method has turned its attention to this subject matter and taken the lead in answering these timeless mysteries.

So, in spite of an unsavory early dalliance with philosophy (hey, I was young, and everyone experimented back in those days), I became a scientist. I wanted to know the answer to such questions as how the universe came into existence or why the laws of nature are the way they are, and it was clear that the study of science was the only credible path forward.

As I grew older, I came to realize that I was not the first science-minded individual to have a similar interest. Indeed, Einstein spent the latter half of his career devoted to studying the same questions. Scientists even have a single proposed name for the answer to all of those puzzles—we call it a "theory of everything."

The name is grand, and it suits the goal. Physicists aspire to develop a theory that literally answers all questions. (Well, ones dealing with the nature of reality at least. Nobody really expects to understand perplexing mysteries like why people put ketchup on hotdogs. But I digress.) Unlike the answer to most questions, which inevitably generates another "but why is that?" question, a proper theory of everything will allow no further questions. The theory will be self-evident.

Admittedly, we cannot be sure that a theory of everything exists. But it is a lovely goal and people will continue to strive for it. Indeed, there are some people who claim that a well-developed theory of everything is right around the corner—something that we can hope to see published in the next few years or decades.

And that's what this book is about. I have spent my professional life doing physics research at the very cutting edge of what we know—venturing from the shores of the known, out into the oceans of our ignorance. Thus, I have some familiarity with the subject matter and some experience with the pace at which discoveries are made.

It is my opinion that rather than being in the position where the announcement of a correct theory of everything is imminent, we are in the situation where we still have a long way to go. Perhaps this could be viewed as an attempt to pour cold water on the enthusiasm that drives an entire field of scientific inquiry, but I prefer to think of it as adding a dose of reality to the conversation. I think it is important to have a clear vision as we move forward.

In this book, I will tell you about our current theoretical understanding of the laws of the universe. Two very successful theories—the standard model that describes the quantum world and Einstein's theory of general relativity that governs the cosmos—give us an enormously comprehensive understanding of the laws of reality. Together, they truly are a triumph of scientific achievement.

However, the two models are incomplete, so I then turn my attention to recent and ongoing attempts to extend our current knowledge and develop a theory of everything. These attempts span a century of hard work by brilliant minds and, sadly, none have been successful. I will then explain why current attempts to devise a theory of everything are doomed to failure—scientific hubris, writ large.

While this might seem to be a depressing place to arrive at the midpoint of a book, this is a hasty characterization. I can say this because professional researchers know, in broad strokes, how to move ahead. The universe has shared with us breadcrumbs that mark a trail forward. In fact, there are many trails that lay before us, and following each of them will bring us closer and closer to our goal. This book will describe in real and concrete detail the mysteries that lie ahead and present the myriad of solutions that might bring a theory of everything closer to reality.

In short, this book gives you a real insight into the true status of the ultimate explorations of the laws of nature. And, when you reach the end, you will have a deep appreciation of the research frontier that will occupy the professional scientific community for the next few decades.

Of course, I cannot take credit for the information found in these pages. No single person or even groups of people can. Indeed, the path toward a theory of everything is possible because of the tireless efforts of thousands of men and women, past, present, and future, who have devoted their lives to studying the universe's mysteries. I tip my hat to each and every one of them.

I would like to thank my family, who gave me the time and space to pen these words and share with you this most gripping tale. And I must certainly thank the people who read early drafts of this manuscript. This book has been immeasurably improved by the comments of Cherie Bornhorst, Keith Calkins, Mike Fetsko, Jeff Funkhouser, Eduardo Márquez, Kevin Martz, Dee Dee Messer,

Rebecca Messer, Frank Norton, Mike Plucinski, Rebekah Randall, and Susan Wetzler. I remain in their debt.

Finally, I would like to thank you, the reader. I hope that you will find the journey as exciting as I do, and you will come to understand how science will move forward so we will one day know the answers to the grandest questions of all.

CHAPTER 1

GOD'S THOUGHTS

In 1922, a young woman by the name of Esther Polianowski came to Berlin. Born at the turn of the century in Zhytomyr, Ukraine, she left her homeland in 1919 to escape civil war and the persecution of Jews that was common at that time. Following a series of harrowing and dangerous adventures, she joined a group of pioneers in Palestine. After a brief return to the Ukraine to help her widowed mother and siblings escape to live in safety with her pioneer friends, Esther decided to go to Germany to pursue her dream, which was the study of physics.

The 1920s in Germany was an exciting time for a physics student. The development of quantum mechanics beckoned on the horizon, and the discussions in university seminars were both interesting and passionate. Ms. Polianowski first met Albert Einstein when she entered the University of Berlin, but she began to see more of him when she became a third-year student.

In 1925, the rising antisemitism in Germany began to trouble her and she went to talk to Einstein at his house, to seek his advice. They went for a walk, where he suggested perhaps going to Holland to study and, when she said she could not afford to go there, he suggested that she go to England to work at the Cavendish Laboratory at the University of Cambridge.

Einstein's Unfinished Dream. Don Lincoln, Oxford University Press. © Oxford University Press 2023.
DOI: 10.1093/oso/9780197638033.003.0001

She eventually took his advice, but it is her account of that leisurely stroll that gave us what was perhaps one of Einstein's most poetic phrases. Shortly after Einstein's death in 1955, she published this account under her married name of Esther Salaman in *The Listener*, a weekly magazine published by the BBC.

Their conversation ranged far and wide, from his most recent efforts to unify the theories of electromagnetism and gravity, to German and Russian literature, from Dostoevsky to Goethe and Faust. They spoke of philosophy and Einstein's opinion of the discipline's insubstantial contribution to science.

But it was when Esther expressed an interest in learning French, so she could read French literature as it had been written, that Einstein revealed his deepest goals as a scientist and great thinker. He dismissed traveling to France, because he wasn't much of a people person. His interests were mostly about his research—his desire to understand the rules of the universe. He said, "I want my peace. I want to know how God created this world. I'm not interested in this or that phenomenon, in the spectrum of this or that element. I want to know His thoughts; the rest are details."

Now, that phrase, "God's thoughts," was never meant to be taken literally as in a literal deity with literal thoughts about an initial explicit intent for the inner workings of the universe and how the cosmos came into existence. Einstein didn't believe in that kind of god. Instead, he believed in nature and the majestic natural processes that govern it. Although he often spoke in a way that could be interpreted as being religious, he was a strict scientist who believed that it was possible to understand the laws of nature.

"God's thoughts" should be considered a beautiful metaphor, one which anyone can appreciate—even the most adamant atheist.

The phrase should be understood in the way it was intended, which is that Einstein wasn't interested in petty minutia, such as why some elements conducted electricity, while others don't. He wanted to understand much deeper questions. He wanted to understand just *why* the universe is the way it is. He was talking about the most fundamental underlying rules from which all other rules derive.

And this grand goal brings us to why you're reading this book. The phenomenon which Einstein called "God's thoughts" has a modern name, although admittedly one with a far less poetic flair. Modern scientists instead talk about an ultimate theory—one which can answer all questions. And, avoiding the whimsy of modern physics, which talks of quarks and color and flavor, scientists call this ultimate goal of science "a theory of everything," or TOE. You might have heard of another, similar phrase, a grand unified theory, or GUT. The two ideas are quite different, and we'll eventually learn about the distinction.

What is a theory of everything? It is, quite literally, a theory of everything—a theory which hopes to explain all phenomena—everything. So just how do scientists imagine such a theory might look?

Well, the first thing is that the theory must be simple. Note that I don't mean simple as in being so easy that everyone can easily understand it. That would run in direct conflict with our experience, where ideas like quantum mechanics, relativity, and $E = mc^2$ are all a bit mind-bending. No, I mean something else by the word "simple." I mean that ultimately and at the deepest and most fundamental level, the universe is constructed of just a few building blocks (and maybe only one). Building blocks that don't interact would result in a very boring cosmos indeed, and we know that we don't live in such a universe. In our world, objects interact with one another by what appears to be a variety of forces.

But we have learned that phenomena that look quite different can have a common origin, much like we know that zebras and humans both arose from a common ancestor. Lightning and the seemingly unrelated behavior of a compass can both be explained by a single theory.

Now you may have heard that scientists believe that a theory of everything is out there, waiting to be found. But you might not have understood exactly why it seems so evident to researchers that the universe has a simple set of rules that govern it. So let's talk about a handful of seemingly very different phenomena and try to dig down to their ultimate explanations. We'll pick five: a fizzy soda, a kitten, a volcano, an orbiting planet, and a supernova. These five phenomena appear to be very, very different, but we shall see that they all have common origins when you delve deeply enough.

To pursue this goal, we must channel our inner two-year-old. Every answer must be met by the question "why?" By doing so, we will dig down through layer upon layer of cause and effect, hopefully finding a final origin that has no cause and, therefore, no further "why?" questions are required. So let's begin.

Why is a soda fizzy? Well, it's fizzy because it's liquid, and it contains carbon dioxide gas suspended in it. So there are two phenomena: the phenomenon of liquid and the phenomenon of gas. Why is the carbon dioxide gaseous at room temperature and yet the surrounding water is liquid? To answer that, you need to invoke the properties of molecules—the molecules of carbon dioxide and water. Water molecules are bent and look a little like a "V." In contrast, carbon dioxide is straight, more like the letter

"I." This, and the details of the bonds connecting the atoms in the respective molecules, gives the two substances different properties. Water molecules are said to be polar, which leads to many of water's unique properties. In contrast, carbon dioxide is not polar. Taking all of these factors into account, water is liquid and carbon dioxide is gaseous at temperatures comfortable to humans.

In the examples we are exploring, I will not talk in detail about the relevant phenomena. While fascinating, it would derail the main point, which is to get at the ultimate causes. So, rather than talking about the finer nuances of these molecules, we might ask, "What is it that gives the molecules these properties?" The answer is, of course, the atoms of which the molecules are composed.

Carbon dioxide is made of carbon and oxygen, while water is made of hydrogen and oxygen. The atoms of each of these elements are composed of an atomic nucleus, surrounded by electrons. Carbon hosts six electrons, oxygen eight, and hydrogen only one. The electrons are bound to the atomic nuclei governed by the force of electromagnetism, which provides a force of attraction between nuclei and electrons. And the way in which the electrons and nucleus interact is governed by the principles of quantum mechanics. Thus, the next question becomes "Why is electromagnetism the way it is, and why does quantum mechanics work the way it does?" Restricting ourselves to the phenomena we can investigate by studying fizzy soda, we can't really answer those questions; but we can say with some confidence that electromagnetism and quantum mechanics are important components of what Einstein so poetically called "God's thoughts."

So let's now turn our attention to kittens. Kittens really seem to be about as different from fizzy soda as you can imagine, unless you include kitten burps (which I'm not even sure is really a thing). But what questions can you ask about kittens? Well, you might ask, "Why they are fuzzy?" The answer for that is easy: It's to keep them warm. Channeling our inner toddler, we can ask, "Why do they need to be warm?" And the answer to that is that having a stable temperature allows them to regulate their metabolism. The reason they need to have a controlled metabolism is because it allows their life to go on. Too slow a metabolism and a kitten would barely be able to move—a tiny torpid tiger, so to speak. Metabolizing too quickly means that the kitten's energy comes out all at once and it would spontaneously combust. So measured and controlled metabolism is key.

So what controls the metabolism? Chemistry. Very, very, complex chemistry. And what governs chemistry? Molecules and atoms. And, with that, we have returned to the same place we arrived as we considered fizzy soda. The reason that kittens can exist also boils down to the laws of electromagnetism and quantum mechanics.

So is that it? Is the answer to why the universe is the way it is just a matter of electromagnetism and quantum mechanics? It's certainly a piece of the story, but there's more.

To expand our list of fundamental origins, let's talk about a volcano. Certainly, a volcano is quite different from both a fizzy soda and a kitten, but it has some commonalities. A volcano emits lava, which is molten rock. Molten rock is liquid, like the water in the fizzy soda. Rock has a different chemistry than the type we've discussed already, but rock is also made ultimately of atoms with their quantum and electromagnetic rules. But there's another facet here that those two principles don't explain. Why is the rock hot?

The Earth was hot when it formed through the bombardment of small, asteroid-like bodies. Slam two things together hard enough and they'll get hot. (Don't believe me? Take a hammer and continuously hit a rock for five minutes and then feel the hammer. What do you get? A hot hammer.)

But the Earth was made a long time ago and that heat should have dissipated by now. So where does the heat come from? Do quantum mechanics and electromagnetism explain it?

The answer is no. It turns out that the origin of the Earth's heat comes from the decay of the nucleus of radioactive atoms, mostly potassium, uranium, and thorium. And radioactivity is not governed solely by quantum mechanics and electromagnetism. For that, you need to look inside the nucleus of atoms, which is something we completely ignored when talking about chemistry (and soda and kittens).

Nuclear decay is the process whereby atomic nuclei break apart or emit a variety of different kinds of particles. Atomic nuclei are made of a mix of protons and neutrons. Protons are positively charged, while neutrons are neutral, and they are held together very tightly in the nucleus.

So, considering the atomic nucleus using the microscope of quantum mechanics and electromagnetism, we immediately see that we must have missed something. Electromagnetism has the property that electric charges with opposite sign attract one another (like the negatively charged electron and the positively charged proton), while electrically charged objects with the same sign repel one another (like the protons in the nucleus of atoms). Accordingly, if only electromagnetism existed, we'd never see atomic nuclei. The protons would blast themselves apart and—voila!—no nucleus. In addition, neutrons have no electrical charge, so they don't experience electromagnetism, and yet they are bound deeply inside nuclei.

With the discovery of the proton in 1919 and the neutron in 1932, it was clear that there must be another force (unimaginatively called the strong nuclear force) that holds the protons and neutrons together. It is both different in how it works and much stronger than electromagnetism. It turns out that the strength of the strong force inside the nucleus is between a hundred and a thousand times stronger than electromagnetism in that same environment.

There is a specific type of decay that affects the thorium and uranium that heats the Earth. This decay occurs when those nuclei break apart, emitting the equivalent of the nucleus of a helium atom, and it is called alpha decay. In the field of radioactivity, the name for an emitted helium nucleus is an alpha particle, and it contains two protons and two neutrons. When the alpha particle is ejected from uranium or thorium, the remaining nucleus becomes a different element. And, with the story of volcanoes, we add another important fundamental guiding principle of the universe—the strong nuclear force.

I said that potassium was also a radioactive element that contributed to the heat that melts rock. It is governed by the three fundamental principles we've uncovered so far, but the radioactivity is of a different type. Potassium emits an electron when it decays. None of electromagnetism, quantum mechanics, or the strong nuclear force we've encountered can explain the emission of an electron by a nucleus. So physicists had to postulate yet another subatomic force, this one called the weak nuclear force. The weak nuclear force is, rather unsurprisingly, much weaker than the strong force. We'll discuss it in more detail in the next chapter, but we must add that to the three other known fundamental phenomena. We're up to four.

Before we explore our other two seemingly unrelated macroscopic phenomena (orbiting planets and supernovae), it is perhaps important to pause for a moment and take stock. As we have investigated very different things in the world around us, we've found that very disparate phenomena turn out to arise from a small number of smaller objects, governed by a few underlying principles. Soda, kittens, and volcanoes (and many, many more things) can be explained by four causes. The path from complex to simple is a well-trod one.

Now let's add the orbiting of a planet. What makes a planet orbit around the Sun? While it's true that the planet is governed by the four underlying causes we've encountered, none of them actually answers the question. To explain orbital motion, we need to add another phenomenon to our tool chest. We need to include gravity. It's unfortunate that another fundamental force and principle is needed, but it's just one more. And gravity explains phenomena beyond the orbit of planets. It explains why we don't float off the Earth. It explains how galaxies move and rotate. It explains the stately march of a comet as it falls toward the Sun for a visit, before it careens off into deep space. Gravity is another important and core governing principle. We're up to five.

Our last seemingly unconnected thing we're going to explore is a supernova, which occurs when a massive star uses up its nuclear fuel and explodes.

There are a few classes of supernovae, but we'll just pick one for illustration. A massive star—far more massive than our own sun—is formed from a cloud of hydrogen gas. Hydrogen gas is made of molecules, and we've discussed what governs molecules. As the gas accumulates and compresses, it gets hotter and hotter. Gravity is what pulls the gas together, and the principles of quantum mechanics and electromagnetism explain what makes it hot.

The temperature of the star increases until, eventually, nuclear physics takes over. The same basic principles that make the core of the Earth hot become important, as the strong and weak nuclear forces cause hydrogen nuclei to fuse into heavier nuclei, giving off even more heat. The weak nuclear force even turns some protons into neutrons, so heavier and heavier elements can be formed. This process goes on for many millions and frequently billions of years. The hydrogen is slowly converted into heavier and heavier elements until none is left. This sets into motion other nuclear reactions, making heavier and heavier elements—a process that alters the temperature of the center of the star.

However, one day, all possible nuclear reactions have occurred. No more energy can be extracted in this way, which means the heat (and resultant pressure) that balanced gravity's inexorable grip eventually disappears. Gravity pulls the remaining matter of the star into a tiny size, which drives up the temperature so high that the pressure overcomes gravity and blasts the mass of the star across the universe and announces its death to the cosmos.

To explain a supernova requires all of our fundamental principles: quantum mechanics, electromagnetism, the strong and weak nuclear forces, and gravity. But, coming full circle, I must remind you that we started this with a discussion of soda and kittens. The destruction of a star can be explained by the same origins as the adorable ball of fluff that purrs in your lap. Even though I've known this for decades, that observation continues to be completely amazing to me.

It's perhaps worth mentioning that the supernova is connected to our lives in another way that we may not fully appreciate.

Without stars, the universe would consist of about 75% hydrogen, and 25% helium, with basically nothing else. It is in the kiln of a star that heavier elements are forged and in the blast of a supernova that both adds to the richness of the elements in the universe and spreads them across the cosmos. Without that process, we would not exist. It is easy to see why Carl Sagan often noted that we are all star stuff.

With our journey into the ultimate origins of a handful of things that seem very different, indeed perhaps completely unrelated, we have come to a very significant conclusion. The complexity of the universe arises from a very few progenitor causes. In this introductory chapter, we have identified a few. There are the protons, neutrons, and electrons that comprise atomic matter. If we add the governing rules of electromagnetism, gravity, and the nuclear forces and toss in some quantum mechanics, we can make up essentially everything we've ever observed.

And that is essentially what Einstein meant when he said he was interested in God's thoughts. He wanted to know the deepest and most fundamental rules, not the details.

Our understanding of the universe has evolved since Einstein's time. We know that the protons and neutrons are made of smaller particles still, called quarks. We have a much better understanding of how the forces actually work, both on a subatomic scale and writ large, across all of creation. In the next chapter, we'll talk about a modern understanding of the world—both subatomic and cosmic. We'll spend some time talking about how we have worked out connections between phenomena that seemed different and how forces that seemed dissimilar are the same. Essentially, the next chapter will bring us up to speed on what modern research scientists consider to be the best answer to Einstein's dream and why expectations for a speedy discovery of a theory of everything are optimistic. And then, in the following chapters, we'll dive off into the unknown, trying to push back our frontiers of ignorance, and learn of current attempts to understand the universe around us at a deeper and more fundamental level. In short, we'll learn just how researchers are taking step after inexorable step toward a theory of everything, trying to answer the deepest question of all: Why are things the way they are?

CHAPTER 2

CURRENT KNOWLEDGE

The search for a theory of everything goes way back—long before the invention of writing—and theories existed in many cultures. The early ones weren't cast in a scientific way; they were more religion than science, and they incorporated other topics as well. But questions about how the universe came to be and why the world is the way it is are deeply embedded in the human psyche. The questions this book explores aren't new ones.

Probably the first serious attempt to approach the problem in a modern way and to try to get a handle on the building blocks of the cosmos began with the Greek philosopher Empedocles. He was born in 494 BCE and was interested in the same things that bother physicists today. Essentially, he wanted to know the ingredients of the universe. He postulated five: the traditional Earth, Water, Air, and Fire, as well as a fifth ingredient that we would now call the void. Combine the traditional four elements in the right mixtures, and you could make ordinary matter. That matter would exist in the void.

The Grecian tradition of merging mathematics and physics also originated about the same time. Plato was born some sixty or so years after Empedocles, and he was fascinated with geometry. He explored the mathematics of what we now call Platonic

Einstein's Unfinished Dream. Don Lincoln, Oxford University Press. © Oxford University Press 2023.
DOI: 10.1093/oso/9780197638033.003.0002

solids—three-dimensional shapes whose faces are all identical. Examples are four-sided pyramids, which have a triangular face, or cubes, with each face a square. If you are a Dungeons and Dragons player, your dice are examples of Platonic solids.

It turns out that there are five possible Platonic solids: the tetrahedron with four faces; the cube with six; the octahedron with eight; the dodecahedron with 12; and the icosahedron with twenty. In Plato's dialogue *Timaeus*, he assigned the five Platonic solids to the five elements first discussed by Empedocles. While the two men promoted a theory of matter that didn't turn out to be correct, Plato's work is an early example of the melding of science and mathematics that has proven to be such an effective tool over the centuries.

Now Empedocles wasn't the only person thinking about the nature of matter. Democritus was born about thirty years after Empedocles, and thus about thirty years before Plato. He was a member of what is called the atomist school of ancient Greek philosophy. Atomism proposed that matter was made of very small objects—too small to see with the human eye. These objects could not be broken into smaller things, and they were called *atomos*, which means "uncuttable" in Greek. Democritus imagined that his *atomos* were the building blocks of reality. Philosophically, Democritus would have been very comfortable with the modern quest for the theory of everything.

However, Democritus's ideas, while insightful in arguing for the existence of tiny building blocks, were wrong in detail. He imagined an enormous number of *atomos* of various shapes, with properties he deduced from how we experienced the materials of day-to-day life. There were pointy ones, to be found in lemon juice, for example, smooth ones to be found in olive oil, and so on.

Some atoms were hard, and some were squishy. While his ideas might seem quaint by modern standards, the idea that matter was made of a handful of building blocks remains credible even today.

There is a long path from Democritus's era to the present, trod by countless individuals who wanted to understand the ultimate nature of matter. Legendary figures from chemistry and physics delved deeper and deeper into the question, resulting in knowledge about a series of ever-smaller objects, from molecules to atoms to protons, neutrons, and electrons, to now even smaller objects, which scientists have discovered in the last half a century or so.

While the history of that journey is a fascinating one, and there are several excellent references in the Suggested Reading that will fascinate those interested in the history of physics, that history is a mere sidebar to the topic we're interested in, which is a possible future theory of everything. In order to move forward, we must focus on what we know now and forgo tales of how we got here. So, that said, what do we know?

Quantum Particles

If humanity has spent millennia trying to work out the ultimate nature of matter, surely we must know the answer by now, right? Well – no – we don't. But we do know a lot. In fact, we have not one, but two, theories which, when taken together, explain an awful lot about the universe. They are called the standard model of particle physics and Einstein's theory of general relativity. We can think of the two theoretical constructs as prototype theories of everything. The first one describes the quantum world, while the second describes the cosmos.

So just what are these two theories in detail? What do they do? How do they interact? Do they interact? And just how close to a theory of everything are they? Let's find out.

We'll begin with the standard model of particle physics. What is it?

Let's start out slowly and begin with the familiar—molecules. Molecules make up every object you've ever seen. If you look at the ocean, with all of its mercurial nature, what you're seeing is a huge and liquid pool of molecules of dihydrogen monoxide or, to chemists, H_2O. Ordinary table sugar? It's called sucrose or, to professionals, the mouthful "(2R,3R,4S,5S,6R)-2-[(2S,3S,4S,5R)-3,4-dihydroxy-2,5-bis(hydroxymethyl)oxolan-2-yl]oxy-6-(hydroxymethyl)oxane-3,4,5-triol." Admittedly, the technical name doesn't naturally roll off the tongue, and it would be unwieldy for your grocery shopping list, when your sugar bowl runs low. Sugar, that tasty substance, is made of molecules with a chemical formula $C_{12}H_{22}O_{11}$.

In the search for a theory of everything, molecules are a huge improvement over earlier thinking, but they are pretty big by modern standards, and there are a whole bunch of them. They are definitely not the ultimate building blocks. The size of molecules varies depending on how complicated they are, but simple molecules have a size in the ballpark of about a billionth of a meter or so, and they are made of even smaller things. So what's the next step in our search for the ultimate building blocks?

Those chemical formulae for water (H_2O) and sucrose ($C_{12}H_{22}O_{11}$) are telling us something; indeed, they are a code for those who have a basic understanding of chemistry. The letters indicate a smaller building block of molecules. These smaller building blocks are called atoms.

Atoms can rightfully be considered to be the building blocks of the chemical world. And there are lots of varieties of atoms—nearly a hundred found in nature. Each atom is the smallest instance of what is called an element. Elements get their name from the word *elemental*, implying that they cannot be changed and that they are ingredients used to build the cosmos. One can think of atoms as like Legos, with the atom of each element corresponding to a different Lego piece. And, if you have enough Legos of different kinds, you can make just about anything. (Seriously... look online for "Lego sculptures." They can be amazing.) Similarly, every object you've ever seen is composed of simply the right mix of atoms of a small number of elements.

The elements have names and symbols, like hydrogen (H), helium (He), carbon (C), oxygen (O), gold (Au), and so on. You know what gold looks like—a metal that might have made up a ring you've seen on your parent's finger. However, if you could zoom in on a gold ring with a super powerful microscope, you'd eventually see individual gold atoms. If you need a visual to help you imagine it, consider a sand dune and a grain of sand. The sand dune could metaphorically be a pile of some element, while the grains of sand are individual atoms. Of course, sand isn't an element, but the analogy conveys the right basic idea.

The late 1800s were a great time for people trying to work out a theory of everything. While scientists have generally imagined that there are but a few building blocks of matter, chemists had found nearly a hundred. A hundred is a lot, but it was a big improvement over how people viewed matter before the invention of chemistry. And scientists were able to work out rules determining how the elements interacted with one another. It was a huge step forward to know that water was made of two hydrogen atoms and one of oxygen and to realize that ice, liquid water, and steam were the same thing. Furthermore, all other substances were just different combinations of those hundred or so chemical elements.

Of course, scientists are a curious sort, and they continue to look into things, sometimes with surprising results that can lead to an advance in knowledge. By looking at atoms, researchers found out that they weren't truly elemental and that these atoms contained even smaller building blocks. They discovered that atoms contained electrons in 1897, that they contained protons in 1919, and that they contained neutrons in 1932.

With the discovery of protons, neutrons, and electrons, physicists rejoiced. Instead of one hundred chemical elements, that somewhat unwieldy collection of elements could be created by the right mix of these three smaller particles. For instance, hydrogen consists of one proton and one electron. Oxygen has eight protons, eight neutrons, and eight electrons. The replacement of nominally one hundred building blocks with only three smaller ones was another tremendous simplification and suggested that scientists were closing in on a single particle that all matter was made of.

While a lot of progress had been made in the first half of the twentieth century in the search for the ultimate building block, there were other mysteries. For instance, there seemed to be some kind of radiation that came from space and that needed an explanation. Researchers call this type of radiation "cosmic rays." And the study of cosmic rays leads us to the next story in this tale.

The nomenclature of cosmic rays is a bit confusing. Originally, the term "cosmic ray" was used to indicate the radiation that early physicists were investigating near the Earth's surface. However, they didn't know the nature and origin of cosmic rays. Working out what was going on took decades, and the history is rich and interesting, but we can skip ahead to a modern understanding of the phenomenon.

The Earth is constantly pummeled by fast-moving subatomic particles that fill the cosmos. Most of these particles are simply protons emitted from stars, supernovae, or other interesting cosmic events. And, as I mentioned, the nomenclature of cosmic rays is a bit confusing. While it's common to call the radiation detected at the Earth's surface "cosmic rays," people also refer to the protons hitting the atmosphere by the same name. However, as we will soon see, the protons never make it to the Earth's surface. Still, the protons are the origin of the radiation observed by early researchers, so there is a connection. The bottom line is that the phrase "cosmic rays" can refer both to the high-energy protons in space or the radiation observed in detectors located on the ground. You need to know the context to understand which a writer or speaker is talking about. I will try to be clear when I mention them in this text.

Protons from outer space travel long distances, and they smash into the Earth's atmosphere. The atmosphere is made of atoms of nitrogen, oxygen, and the like. And those atoms—like all atoms—are made of protons, neutrons, and electrons. So, when a high-energy proton from space plows into the atmosphere, the result is a collision between the proton and one of the three constituents of atoms. For the proton, it's a cataclysmic quantum catastrophe.

When high-energy protons collide, what happens is very different from what happens when two cars collide. And, like many physics phenomena discovered in the last century or so, we need Einstein to help us understand what is going on.

Einstein's equation $E = mc^2$ is a famous one. Formally, it says that energy (E) is equal (=) to mass (m), times a large constant (c^2). Practically, it means that energy can convert into mass (e.g., matter) and back again. And that opens the door to some fascinating behavior.

So what happens? The proton smashes into some atom that is present in air. Say it hits another proton. The proton from space is moving very fast—often near the speed of light. Sadly, neither proton survives the collision. Metaphorically, they are shattered. However, here is where interesting things happen. The huge energy that the proton from space had when it was traveling so fast is converted into other subatomic particles. Now, not all of the energy is converted into particles, meaning that these daughter particles also move downward toward the Earth at high velocities. While each collision is different in detail, this process turns two particles (the proton from space and a stationary proton on Earth) into (say) eight particles of lower energy. The eight is just an example—different collisions result in a different number of daughter particles. A representative cosmic ray shower is shown in Figure 2.1.

For illustration purposes, we'll stick with the representative eight particles. Each of them then smashes into another air molecule, and the process repeats itself. If each collision results in eight particles (and it doesn't in reality—the actual number varies a lot), the number of particles grows from 2 to 8 to 64 to 512, and so on. After each collision, the daughter particles have less and less energy, and eventually this energy gets low enough that it is no longer possible to make more daughter particles. This "shower" of great, great, (many times great), granddaughters of the original

Figure 2.1 Cosmic rays from space (mostly protons) smash into the Earth's atmosphere and set up a cascade of particles that pass through the atmosphere. Confusingly, these secondary particles are also often called cosmic rays. (Figure courtesy of CERN.)

two protons then hit the Earth's surface. These are the particles that puzzled early scientists.

I was vague on exactly which particles were created in the various collisions. That's because I haven't introduced them yet. In fact, there are dozens and dozens of different kinds of particles that we'll learn about soon. Each collision in the atmosphere is generally far more complex than the simple process I've outlined here.

In any event, in the days before particle accelerators, these cosmic rays at the Earth's surface were the highest energy particles that were accessible to researchers. The scientists expected these particles would help them explore the nature of Einstein's theory of special relativity, which governs objects moving quickly, and quantum mechanics, which explores the world of the small. And they were successful; however, some surprises arose.

For instance, in 1932, an American scientist by the name of Carl Anderson used cosmic rays to discover a particle that seemed to be essentially an electron, but with the opposite electrical charge. While an electron has a negative charge, this new particle had a positive one, which led to the particle's name—the positron. It turns out that Anderson had discovered a substance called antimatter.

Antimatter had been predicted a few years prior when British theoretician Paul Dirac had combined the laws of quantum mechanics and Einstein's theory of special relativity. The result, called relativistic quantum mechanics, described the behavior of small particles moving very fast. One of the consequences of this theory is that it predicted the existence of "antimatter," which is essentially the opposite of familiar matter. Although it was not obvious at the time, Dirac's theory predicted (and subsequent discoveries have verified) that the electron was not the only subatomic particle with an antimatter counterpart. There was also the antiproton, antineutron, and even anti-equivalents of the particles within protons and neutrons, which we'll encounter soon.

Antimatter and matter annihilate when they touch, and the result is a huge amount of energy. For instance, if one were to take a gram of antimatter—say the amount in a mid-sized paperclip—and combine it with a gram of matter, the energy release would be approximately equivalent to the bomb blasts at Hiroshima or Nagasaki.

The prediction and discovery of the existence of antimatter might be the key moments when particle physics split off from chemistry, atomic, and nuclear physics. The positron was a particle that is not a component of ordinary matter. It is only created when huge amounts of energy are present. With the discovery of antimatter, physicists ventured into the unknown, without any guidance from the classical sciences.

For their contribution to the science of antimatter, both Dirac and Anderson received Nobel Prizes, but Anderson wasn't done. In 1936, he discovered another particle that nobody imagined existed. Using the same techniques he employed to discover the positron, he discovered a particle that had many of the properties of the electron, but it was about two hundred times heavier and it decayed in about a millionth of a second. This particle has had various names in the past, but the scientific community has settled on the term "muon." The name has no real significance and simply originates from the custom at the time of naming particles with Greek letters. The symbol for this new particle was the Greek letter mu (μ), and the "on" shadowed the ending of familiar particles, like the proton, neutron, and electron.

The muon was completely unexpected—no hints in chemistry suggested that it existed. In fact, when quantum mechanics legend I. I. Rabi first learned of the existence of the muon, he is reported to have exclaimed, "Who ordered that?" The muon seemed to have no role at all in the behavior of ordinary matter. It was just a particle that researchers could eventually create at will, but it decayed too rapidly to participate in our daily life. It truly was a mystery.

There is one more discovery we should highlight in our truly whirlwind detour into the history of particle physics. This one is the ghost of the quantum world. It is called the neutrino.

In the late 1920s, there was an epic conundrum in the physics community. Researchers studying a rare form of carbon atomic nuclei saw that the nuclei would occasionally decay. Specifically, a neutron would decay into a proton and electron. At first blush, this seems just fine. A neutron is electrically neutral, with zero charge. A proton has +1 charge, and an electron has −1 charge. So, after the decay, the two particles had no net electrical charge, just like before the decay (e.g., $+1 + (-1) = 0$).

Furthermore, the energy budget of the decay was somewhat sensible. Since $E = mc^2$, one can talk of energy and mass interchangeably. The mass (or, equivalently, energy) of the neutron is slightly more than the combined mass of the proton and electron. Since the energy of the neutron is higher than the combined energy of the decay products, the decay is allowed from the energy conservation point of view.

However, there were some mysteries. The first thing has to do with energy conservation. Because the energy of the neutron was higher than the daughter particles, the extra energy that didn't go into making the daughters had to go somewhere and that somewhere was in making the electron and proton move. Because the neutron was essentially stationary, it had no momentum. Momentum is one of those quantities that are conserved, so that means that the two daughter particles had to move in exactly opposite directions to cancel each other out.

When you combine the idea that both the energy and momentum of the combination of the proton and electron should be identical to that of the neutron, you can prove that the electron coming out of the decay should have a single specific energy. And, furthermore, all electrons in this form of radioactive decay should have exactly the same energy.

However, when experiments were done, the electron never had the expected value. Instead, the electron energy was a range of values, going from zero up to the predicted value. On the face of it, it seemed that in the quantum world that energy wasn't conserved. That would have greatly shaken the scientific community.

There was another mystery associated with these kinds of decays that is a bit more technical, but no less troubling for scientists. This conundrum has to do with the quantum spin of atomic nuclei and electrons. For instance, in the decay of carbon-14 (e.g., 6 protons and 8 neutrons) into nitrogen-14 (7 protons and 7 neutrons), both the parent and daughter atomic nuclei have an integer amount of spin. However, the electron has a spin of ½. Thus, what researchers observed was that carbon (with integer spin) decayed into nitrogen (integer) plus an electron (half integer). That means the spin can't be the same before and after the decay. While somewhat technical, this was also a serious problem. These two observations seemed to imply that both energy and spin were changed in this kind of radioactive decay. This would mean that the quantum world was way different than the classical world seen in the human-sized world. It was truly confusing.

However, help was on the horizon. In late 1930, Austrian physicist Wolfgang Pauli proposed that perhaps another particle was being emitted when a neutron decayed. This is the particle we now call a neutrino. Neutrinos would have a spin of ½, which would fix the spin problem, and they would interact very weakly, which meant they would escape the detector and carry away energy that was not detected. This noninteraction would solve the missing energy problem. It was a great solution. There was only one problem—no neutrinos had ever been observed.

Given the weakness with which neutrinos interact, detecting them is very difficult. In fact, it took a quarter of a century to demonstrate their existence. It was 1956 when the neutrino was first discovered emanating from a nuclear reactor. Pauli's proposed particle was confirmed.

We should now return to understanding the way in which the standard model is thought to be an important component of a theory of everything. While the neutrino didn't participate in chemical reactions, it was an important part of a common form of radioactive decay and, by the 1930s, scientists had been studying radioactive decay for a long time. So the neutrino should be considered to be a shy member of the family of ordinary matter.

Let's pause and take stock. By 1956, researchers had detected protons, neutrons, electrons, and neutrinos. These are all components of the matter that makes up you and me. They had also discovered the muon, which was like a heavy electron, but didn't fit into ordinary matter in any recognizable way. But it was an important clue that pointed toward a more complicated subatomic world.

In addition to the particles I've mentioned so far, the time period of about 1946 through the early 1960s was a halcyon time for particle physicists. World War II was over and technical advances in radar technology developed during the conflict made it possible to build powerful particle accelerators—devices that would take protons or electrons and hurl them at great speeds at stationary atomic targets.

By exploiting both particle accelerators and (to a decreasing degree) cosmic ray experiments, scientists discovered a veritable zoo of subatomic particles that didn't play any role in chemistry. They could be made by concentrating enough energy into a single spot and turning that energy into the mass of a dizzying variety of particles. And what a diverse assemblage they were. There were particles with a range of values of electric charge, for example −2, −1, 0, +1, and + 2 times the charge of a proton (but never fractional charges). Some had integer spins, and some had half integer spins. Some had masses near or larger than a proton (these were called baryons, coming from the Greek word *barus*, or heavy), while some had masses that were in the range of about a tenth to a half the mass of a proton. These were called mesons, following the Greek word of *mesos*, or middle. There were also super light particles, called leptons, from the Greek word *leptos*, or light. Early on, only electrons and neutrinos were considered leptons, although the meaning of the word *leptons* has morphed over time.

The particle zoo also experienced a variety of forces. We will speak more of forces in the next section, but those particles with electrical charge were governed by the force of electromagnetism. Some particles felt the strong nuclear force, and some decayed via the weak nuclear force. The force of gravity is too weak to participate in subatomic reactions.

The bottom line was that by the early 1960s, many dozens of particles had been discovered in particle accelerator and cosmic ray experiments, with a myriad of properties. Most of them were evanescent, being created from energy, and then decaying rapidly into more familiar particles. Still, dozens and dozens of unknown particles seemed like a step backward for those looking for a simple theory of everything. The days of proton, neutron, and electron appeared idyllic in comparison. Clearly an explanation was needed.

In 1964, two physicists, working independently, had a joint epiphany. One was American physicist Murray Gell-Mann, and the other was Russian-American George Zweig. They hypothesized that the proton and neutron contained within them even smaller particles. We use Gell-Mann's name of "quarks" to label these particles. (In Zweig's paper, he called them "aces.")

In 1964, three quarks were imagined to exist. There was the up quark, with an electric charge of +2/3 that of a proton, and two quarks of charge −1/3, called the down and strange quarks.

Up and down quarks made up ordinary matter. Two ups and one down quark were found in the proton (*uud*, (+2/3) + (+2/3) + (−1/3) = +1), while one up and two downs were found in the neutron (*udd*, (+2/3) + (−1/3) + (−1/3) = 0).

The strange quark was hypothesized to explain a puzzling feature of some of the other particles that had been discovered in the preceding two decades. Some of those particles were easy to make and took a long time to decay, which was . . . well . . . strange. Most particles that are easy to make also decay quickly, and ones that are hard to make decay slowly. Particles that were created easily and decayed slowly were called "strange particles," and the name transferred to the quark. An example of a strange particle is called a lambda-0, Λ^0. It contains one up, one down, and one strange quark. It is slightly heavier than a proton and has zero electric charge, as one can tell by adding up the electrical charges of the constituent quarks. The Λ^0 was one of the many baryons that were discovered in the middle years of the twentieth century.

Quarks were pretty versatile. Take any combination of the three quarks proposed by Gell-Mann and Zweig and you could describe the baryons that researchers had discovered at that time. Furthermore, one could also take any quark and an antimatter quark, and such a combination would make up a meson. That's how they work—three quarks make up a baryon and a quark/antimatter quark pair constitutes a meson.

And all of this was great news for researchers looking for a simple theory of everything. The dozens of baryons and mesons that had cluttered the scientific literature of the 1950s and early 1960s were constructed of a mere three building blocks called quarks. The electron, muon, and neutrino contained no quarks and were their own kind of particle. In the modern parlance, any matter particle not containing a quark is called a lepton.

So that brings us almost to the present. According to modern physics, quarks and leptons are the smallest known particles. I said "almost," because additional quarks have been discovered. Using ever more powerful particle accelerators, researchers have discovered three more quarks: the charm quark (1974, with a +2/3 charge), the bottom quark (1977, with a –1/3 charge), and the top quark (1995, with a +2/3 charge).

The quarks were not the only category of subatomic particles to grow—some more leptons were also discovered. In addition to the electron and muon, the tau lepton was discovered (1975, with a –1 charge). And, rounding out the remaining advances, it became apparent that there were three types of neutrinos, each a cousin to the electrically charged leptons. There was the electron neutrino (1956), muon neutrino (1962), and tau neutrino (2000).

So now we are truly up to date. Modern science knows of three quarks with a charge of +2/3 that of a proton (up, charm, top), three quarks with a charge of –1/3 (down, strange, bottom), three leptons with a charge of –1 (electron, muon, tau), and three leptons with a charge of zero (electron neutrino, muon neutrino, tau neutrino). Twelve quarks and leptons are what are needed to explain all of the forms of ordinary matter that have ever been observed. These are the smallest building blocks discovered so far.

Of course, building blocks are one thing. The rules that govern how they interact and bind together are important as well. For that, we need to explore another facet of reality.

Quantum Forces

In the previous section, we talked of building blocks. However, as anyone who has played with Legos will tell you, there are rules in how they connect. Orient the Legos in some odd direction, and they won't stick together.

For particle physics, it's basically the same. The known forces tell quarks and leptons how to assemble themselves into material that makes up stars, galaxies, and even us. Some configurations are forbidden. So, in order to understand the standard model, we need a modern appreciation of the forces that govern the subatomic realm. That's crucial.

But, in keeping with the goal of this book, we also need to know the story of how certain, seemingly different, forces were first observed and then were shown to have arisen from a single common framework. This process will perhaps give us a road map for future research.

Counting the number of known forces is a bit tricky, as we will see. Most books claim that there are four forces: gravity, electromagnetism, and the strong and weak nuclear forces. I'm not going to discuss gravity here, and we'll have to wait a bit to understand why counting the forces isn't straightforward. Gravity will be covered in the next section, and a discussion of the problem with understanding how many forces exist will come later.

Electromagnetism is the most familiar of the forces that operate in the subatomic realm. But that simple word, *electromagnetism*, hides a fascinating history. Most of us have some experience with electricity and magnetism, but what do we mean by this word, *electromagnetism?*

To answer that question, we need to remember what we know about electricity and magnetism separately. Electricity is, if one might pardon the phrase, shockingly important to understanding our world. We know of lightning, perhaps one of nature's most majestic aerial displays. We certainly know of the electricity that courses through the walls of our houses and apartments and animates much of our domestic modern technology. We know of little shocks caused by static electricity on a dry winter's day. And who can forget Mary Shelley's *Frankenstein*, where a bolt of lightning provided what was called in the eighteenth century "animal electricity" which animated the monster?

In contrast, magnetism seems to be very different. Magnets hold children's art to refrigerators around the world. They are used in junkyards to lift cars to a waiting crusher. They make possible trains that run by magnetic levitation, which are currently found only in Asia. And one cannot forget the intrepid explorer, lost in the wilderness, who was saved by the freely swinging magnetic needle of the humble compass.

Electricity and magnetism long seemed to have absolutely nothing to do with one another, but that misconception was shattered in 1820 when Danish physicist and chemist Hans Christian Oersted discovered that when electric current flowed through a wire, the result was a magnetic field near the wire. He had discovered a connection between these two disparate phenomena.

However, the reverse did not seem to be true. Taking an ordinary household magnet and placing it near a wire didn't cause an electric current.

It was a decade later that British scientist Michael Faraday discovered how to use magnets to cause electricity to flow. In 1831, he found that if he took a coil of wire and moved a magnet in its vicinity, an electric current was set up in the coil. Because the strength of the magnetic field surrounding a magnet depends on how close or distant one is to the magnet, by moving the magnet around, he was changing the magnetic field in the vicinity of the coil. And that was the key concept—while magnetic fields didn't cause electric currents, *changing* magnetic fields did.

The late eighteenth and mid-nineteenth centuries were the heyday of experimenting with both electricity and magnets. Dozens of researchers documented how both phenomena affected an increasingly intricate number of devices. Electricity and magnetism were fascinating and were connected, but the full picture eluded the scientists of the era. Well, until Scottish physicist James Clerk Maxwell turned his attention to the field.

It was in 1855 that Maxwell made his first presentation of a simplified form of what are called Maxwell's equations. These equations inextricably intertwined the phenomena of electricity and magnetism.

He published his work in 1861 in a book called *On the Physical Lines of Force*. His initial formulation was mathematically clunky, comprising twenty equations and twenty unknowns. Maxwell spent years refining his ideas and simplifying his mathematics, including a book called *Treatise on Electricity and Magnetism*, which he published in 1873.

However, it took British mathematician and physicist Oliver Heavyside to boil down Maxwell's insights into the concise form we know today. In 1884, he published four very elegant equations that encapsulate all of Maxwell's work. They are as follows:

$$\vec{\nabla}\cdot\vec{E} = 4\pi\rho$$

$$\vec{\nabla}\cdot\vec{B} = 0$$

$$\vec{\nabla}\times\vec{E} = -\frac{1}{c}\frac{\partial\vec{B}}{\partial t}$$

$$\vec{\nabla}\times\vec{B} = \frac{1}{c}\left(4\pi\vec{J} + \frac{\partial\vec{E}}{\partial t}\right)$$

where E represents the electric field and B represents the magnetic field. J is electric current and ρ is electric charge. c is the speed of light. Those equations might seem a bit daunting, but don't worry; we're only going to use them to point out a few crucial and simple truths.

While all four equations are important, it is the last two that tell the whole story. In both cases, electricity terms are on one side of the equation and magnetic terms are on the other, and they are joined by an equals sign. Literally, Maxwell's equations say that once the appropriate mathematical operations have been done, electricity is equivalent to magnetism and vice versa.

This is an absolutely crucial scientific insight, which cannot be emphasized enough, but I will repeat: *Two phenomena that seem to be completely different both originate from a single, common source.* Magnetism and electricity are the same thing. Even more compelling is that unobtrusive "c" you see in the equations, which is the speed of light. You can manipulate Maxwell's equations and what comes out is the existence of a wave that travels the same speed that light does. Later experiments validated that Maxwell's electromagnetic waves were actually light. In short, this unification brought together not just two, but three phenomena.

This is an example of *unification*, an idea that we will revisit a few times later in this book. Physicists hope that a theory of everything will unify all of the known forces, and we will finally find that all of the distinct and peculiar features of the various known forces are simply different manifestations of a single phenomenon.

To fully drive the point home, consider two blind men trying to understand an elephant. If one is touching the animal's legs and the other is investigating the beast's trunk, the two men will get a very different idea of an elephant. However, a sighted individual can step back and see how the trunk and legs are connected, and her understanding of the two is richer and more complete than that of either of the two blind men.

The idea of force unification is central to the topic of this book, but we also need to know some details describing how the known subatomic forces actually work. So let's turn our attention to that.

Let's start with a big idea. Forces at the subatomic level differ from what we know about the macroscopic world. In the world of planets, people, and penguins, the idea of field is supreme. Here on the surface of the Earth, everywhere you go, there is a gravitational field, and it points more or less toward the center of the planet. This gravitational field is what causes objects to have weight. However, weight requires both gravity and an object. The object feels the effect of the gravitational field and falls downward.

But the field doesn't require the object. That empty spot five feet to your left might have nothing in it, but the gravitational field is there. A purist might point out that the location I just suggested has air in it, so imagine there is instead a large volume that is a perfect vacuum. Gravity is still there. If we put a baseball inside the volume, it would still fall downward, pretty much the same as it would in air. On the Earth's surface, gravity is everywhere.

The field idea also plays a role in classical electromagnetism. In fact, it was in 1849 that English scientist Michael Faraday coined the term to describe what happens when an electrically charged object is placed somewhere, isolated from other charges. Essentially not much happens. However, he noted that if someone put a second electric charge somewhere near the first object, both of them would feel a force. The idea is that the first object set up an electric field, with which the second object interacts. While there are other types of force (e.g., pushing and pulling), within the context of the forces of interest to modern physicists, the classical description invokes a field.

In contrast, at the subatomic level, forces are explained by a quite different paradigm. In the world of the small, forces are caused when matter particles exchange another kind of particle, this one a force-carrying one.

Let's illustrate this idea. Suppose you have two subatomic matter particles that, in the language of classical physics, generate a force that pushes the two apart. At the quantum level, what happens is that one matter particle emits a force-carrying particle. Assuming that the first particle is at rest, when it emits the force-carrying particle, the matter particle must recoil in the opposite direction to balance momentum. Essentially, what we classically think of as a force is simply recoil in the subatomic realm.

Then the force-carrying particle finally makes it to the second matter particle, which absorbs it. Again, conservation of momentum is key. The second matter particle moves in the direction that the force-carrying particle was moving. By exchanging a single force-carrying particle, the two matter particles are pushed apart. That, in a nutshell, is how forces work at the quantum level. Figure 2.2 illustrates this idea.

Each of the known forces has one or more force-carrying particles that mediate that particular force. Those force-mediating particles have a very diverse nature, with different sorts of charges, masses, and so on. But they all cause subatomic matter particles to change their motion. And, since our intuitive understanding of forces also involves changes in motion, that's the connection between the subatomic and the classical worlds.

OK, so now we're ready to talk about the known subatomic forces: electromagnetism, and the strong and weak nuclear forces. We'll talk about gravity in the next section.

Figure 2.2 Forces generated at the quantum level are caused by the exchange of a particle. When one particle emits the force carrier, it recoils. Then, when the other particle absorbs the force carrier, it also recoils. The net effect is a repulsive force. (Figure courtesy of Dan Claes.)

Let's start with electromagnetism, which is the most familiar of the three. In electromagnetism, only objects with electric charge will experience a force. There are two types of electric charge—positive and negative. In the quantum world, the proton is positive, and the electron is negative.

If the two charges are of the same sign (e.g., + + or – –), the two objects will repel one another. If the two objects are of the opposite sign (e.g., + – or – +), they will attract. The strength of the electric force increases as the two objects come close to one another and decreases at large distances. However, the force never becomes zero. In principle, two charged objects will feel a tiny force between them, even if they are separated by galactic distances.

At the quantum level, the particle that transmits the electromagnetic force is the photon. Ordinary light is merely a bath of photons, although photons exist with wavelengths that are not visible. And, because photons don't themselves have electric charge, they don't interact with one another. Furthermore, photons have no mass. That's the gist of electromagnetism.

As soon as one appreciates the nature of the electromagnetic force, it quickly becomes apparent that at least one other force exists. The reason is the following. As I have said, objects with the same sign charge repel one another. We know that, for example, a uranium nucleus contains ninety-two protons, all with positive charge. If electromagnetism were all that existed, the nucleus of all atoms would blow apart.

But they don't. This means that there must be some sort of attractive force that holds atomic nuclei together, and this attractive force is stronger than the repulsive force of electromagnetism. This force is called, rather unimaginatively, the strong nuclear force.

The term "the strong nuclear force" can be a bit confusing, as it historically has been used to describe the force that holds together the protons and neutrons inside the nucleus of the atom, but that concept is a little obsolete. This force that makes atomic nuclei possible is actually just a fortuitous consequence of the modern version of the strong force.

The modern strong force is what binds quarks together inside protons and neutrons. It's very different from electromagnetism. For instance, instead of two kinds of charge (e.g., plus and minus), for the strong force, there are three. Combine these three different charges, and you get a particle with no strong charge.

The name for these three charges is potentially misleading. The name for the strong charge is "color," and the three types are called red, blue, and green. Of course, talking about quarks having color is nonsensical. They have no color in the familiar sense of the word. The name originates from the fact that if you take three lights—red, blue, and green—and aim them at the same spot on a white wall, the place where the lights overlap looks white. Essentially, if you combine these three colored lights, you end up with no color at all. Metaphorically at least, that's identical to adding the three quark colors and getting a particle with no strong charge.

Like electromagnetism, the strong force is transmitted by the exchange of force-carrying particles. These particles are called gluons, because they "glue" the proton or neutron together. Gluons also have no mass.

As I suggested earlier, the strong nuclear force is much stronger than electromagnetism; however, we have to be careful. That's because all forces vary depending on how close two objects are to one another. For the strong nuclear force, objects that are close to one another feel very little force, and the force increases as the objects are separated. If that's confusing, we can make it more concrete by comparing it to electromagnetism. You can think of the behavior of electromagnetism as being similar to what happens when you play with two small magnets—the force decreases as the objects are pulled apart.

In contrast, the strong nuclear force is more like a rubber band or a spring. The force increases as the two ends are separated. However, that trend doesn't continue forever. Once two quarks are separated by a distance of about a femtometer (e.g., a quadrillionth of a meter or 10^{-15} meters), the strong force drops to zero. What happens during that transition is quite complicated—more complicated than we need to understand at the moment—but the bottom line is that the strong nuclear force is near zero when quarks are near one another. The strength of the force increases as the quarks are separated until about a femtometer separation, and then the force again becomes zero.

This behavior sets limits on how big atomic nuclei can be. If they have a radius much bigger than a few femtometers, the strong force drops off, and the infinite-ranged repulsive electromagnetic force takes over. Thus, big nuclei would get blasted apart. So, when we talk about the strength of the various forces, it's important to remember that the comparison depends crucially on the distance between two objects. On the size scale of humans, the strong force is zero.

There is another difference between the strong force and electromagnetism; the gluons themselves have the strong charge. So, unlike the photons, which ignore one another, gluons interact with one another, and this is part of the reason that the strong force acts like a spring.

The final example of the subatomic forces is called the weak nuclear force. It is much weaker than either electromagnetism or the strong force. It is also responsible for certain types of radioactive decay. Like the other subatomic forces, the weak nuclear force has particles that transmit it, and there are three of them: the W^+ with an electric charge the same as a proton; the W^-, with an electric charge the same as an electron; and the Z, which is electrically neutral. All three of them are extremely massive—in the ballpark of the mass of an entire atom of bromine or zirconium.

The W and Z particles both have weak nuclear charge, so they can interact with one another. Furthermore, since both of the W particles have electric charge, this means that they can interact electromagnetically. This hint of connections between electromagnetism and the weak nuclear force is suggestive, and we will revisit that soon.

OK, so let's recap. We've spoken of three forces that are relevant in the subatomic realm: electromagnetism, the strong nuclear force, and the weak nuclear force. They are all mediated by force-carrying particles, and they are all quite well understood. Table 2.1 summarizes their properties.

Table 2.1 Summary of the nature of the known forces that are important in the subatomic realm. The masses are given in units of the proton mass, while the range of the forces is in units of the proton radius. The Higgs field will be introduced later.

Force	Force Particle	Mass	Typical Range	Strength at 1 fm
Strong	Gluon	0	1	1
Electromagnetism	Photon	0	Infinite	10^{-2}
Weak	W & Z	85–97	1/1,000	10^{-5}
Higgs	Higgs boson	133	—	—

The Higgs Boson and Field

We are nearly at the end of our brief description of the standard model, but I've held out a fairly recent development that is a very important, but perhaps unfamiliar, example of a phenomenon that illustrates the goal of unifying the known forces. The most recent unification attempt involves electromagnetism and the weak nuclear force.

Over half a century ago, scientists knew a lot about the subatomic forces, including most of what we've covered so far. But they continued to dig through data and theories, eager to find a key insight that would advance their understanding.

One of the more promising observations was that the weak force was generated through the exchange of W particles, which had electric charge. This tidbit was not the only hint, but it and others led researchers in the early 1960s to devise a theory that combined electromagnetism and the weak nuclear force. These two, very different, forces were replaced by a single force, called the electroweak force. This was broadly analogous to the fusion of electricity and magnetism into electromagnetism.

However, there was a huge problem with electroweak theory—it only worked if all of the subatomic particles had zero mass. And, of course, that was obvious nonsense. Even at that early time, physicists knew that quarks and leptons had mass. So it seems that electroweak theory was stillborn, never to breathe life into the field of theoretical physics.

But electroweak theory was very attractive and compelling. Could this lovely theory be reconciled with the indisputable fact that some subatomic particles did have mass? Well . . . no . . . and yes. Let me explain.

In 1964, three independent research groups presented a series of insights that were encouraging. They proposed that, beyond the physics known at the time, the entire universe was permeated by an energy field that we now call the Higgs field. The name of this field has been a bone of contention among physicists, as it was named after Peter Higgs, one of the individuals in the three research groups. However, he was not alone. There were six men who made up those three groups. Any one of them could have provided their name to this proposed field. However, history can be a fickle beast, and, by chance, Higgs's name was chosen.

The real history is told in a lovely book called *Massive*, by Ian Sample. Details are given in the Suggested Reading. It's a gripping account of a crucial moment in the history of physics in general, and the quest to unify some of the subatomic forces in particular. I strongly recommend this book.

Credit aside, the Higgs field was a solution for the seemingly fatal flaws of electroweak theory. How did this solution work? The Higgs field interacted with the massless particles of electroweak theory and that interaction gave some of those particles mass.

It should be emphasized that the proposed Higgs field did not arise from any fundamental principles. In a sense, one can think of the Higgs proposal as a Band-Aid that patched up a nasty wound in electroweak theory.

If the hypothesis of the Higgs field was right, scientists also expected to find a particle called the Higgs boson. This particle doesn't give mass to the other particles (the field does that), but because the Higgs boson and the Higgs field are connected, the Higgs boson interacts more with heavier particles than lighter ones.

Well, science is science. Just because Higgs and his colleagues predicted the existence of his eponymous field and boson doesn't mean they exist. After all, the history of science is replete with theories and hypotheses that eventually had to be discarded because they didn't agree with data. So the hunt was on for this undiscovered particle.

It took nearly half a century, with lots of fits and starts, and the occasional false sighting, but in 2012, the Higgs boson was discovered. It's kind of like a photon, except that it's very massive. In fact, it's more like a heavier cousin of the very heavy Z particle.

Because the Higgs boson interacts preferentially with heavy subatomic particles, this means that Higgs bosons should interact with other Higgs bosons, although, of this writing these hypothetical interactions have not been directly observed.

While the standard model is not a theory of everything, it is certainly a good step forward, and it represents state-of-the-art current knowledge. But it is perhaps worthwhile to remember our goal, which is to find a fundamental theory from which all known forces derive and with a single (or no more than a few) building block(s). I've given you a fairly complete account of the state-of-the-art thinking about building blocks, but I haven't quite finished the modern story of subatomic forces. Because we don't know the final answer, we have to step back and get the big picture. Figure 2.3 helps us understand the progress we've made.

In the nineteenth century, scientists successfully demonstrated that electricity and magnetism were really a single phenomenon called electromagnetism. And, with the discovery of the Higgs boson, researchers are increasingly confident that electromagnetism and the weak force are actually different manifestations of the electroweak force. Of course, the origin of the Higgs field is still an unanswered question, but we can see some historical precedent as to why researchers think unification is reasonable.

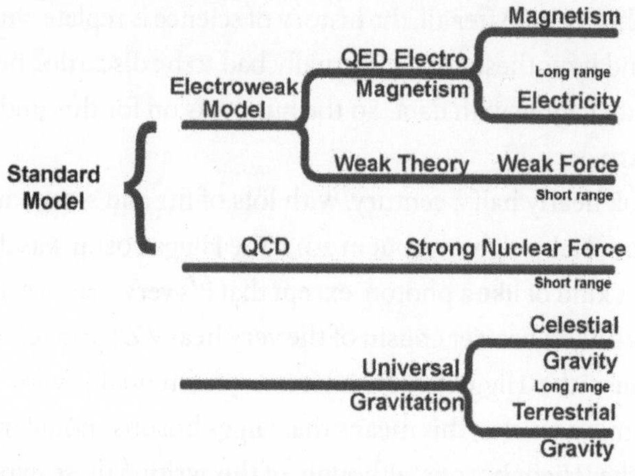

Figure 2.3 This figure illustrates how the various known subatomic forces have been unified. This figure also illustrates how it's difficult to say exactly how many subatomic forces there are. On the right-hand side of the figure, there are six; in the middle, there are four; and on the left, there are only three. And, of course, gravity is merely hypothesized to be relevant in the microcosm. No experiment has demonstrated this conjecture. Thus, one must be very careful when counting the number of known forces. QCD is short for quantum chromodynamics, the modern phrase for the theory of the strong force, and QED is short for quantum electrodynamics, which is the modern theory of electromagnetism.

Of course, we still have the strong nuclear force, which seems to be distinct from the electroweak force, but hope springs eternal, and researchers suspect that further experimentation and theories will show that the electroweak and strong nuclear forces are actually different facets of a combined force, which scientists grandly call "The Grand Unified Theory." That will be discussed in the next chapter. And, of course, we've not discussed gravity yet, and that is the topic of the next section.

With the introduction of the Higgs field, our tour of the standard model is complete. Three forces (strong, electroweak, and Higgs) seem to govern the behavior of ordinary matter. According to state-of-the-art thinking, matter is composed of twelve particles (and the corresponding antimatter ones). Of those twelve, only four of them are key building blocks of matter (up, down, electron, electron neutrino)—the rest are created in the hot conditions of collisions inside particle accelerators, but don't last long. Figure 2.4 illustrates the entirety of the standard model, and the theory is our most current and best theory of the behavior of matter in the microcosm.

But the microcosm is the world of the small. There's a whole huge universe out there, governed by gravity. How does that fit into the picture?

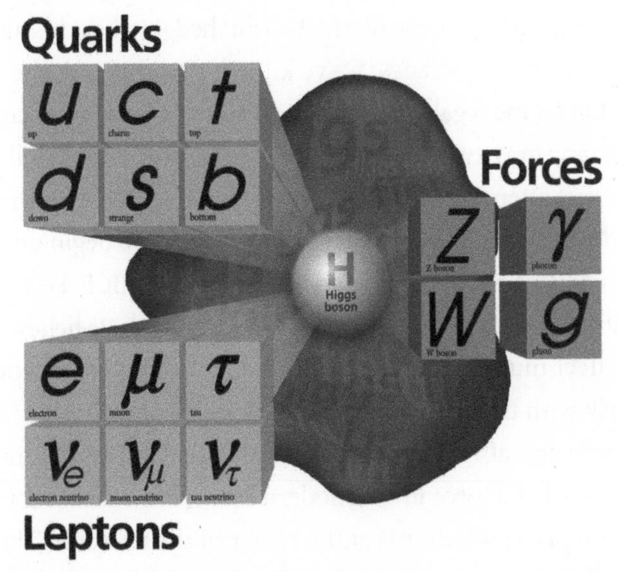

Figure 2.4 The known particles of the standard model. (Figure courtesy of Fermilab.)

Gravity

So far in our discussion of the known forces and particles, I've completely neglected the most obvious of the forces—gravity. This is because scientists have no idea how gravity works in the quantum realm. Now this doesn't mean that we are ignorant about gravity; on the contrary, we know a great deal, from understanding how a ball moves when a child throws it for an expectant puppy to fetch, to explaining the motion of the Andromeda galaxy as at plummets toward a future epic collision with our own Milky Way.

But the way in which we understand gravity is very different from our vision of the subatomic forces. In our theories of gravity, there are no force-carrying particles jumping back and forth between stars and planets. And gravity generally isn't as complicated as the forces we've learned about so far. To a very large degree, gravity is merely an attractive force. I am hedging here because for the last couple of decades, we've known that this isn't the entire story. But let me regale you with the history of how we've come to understand gravity, as there are a few interesting tales to tell.

The first attempts to unravel gravity began, as is often the case, with the ancient Greeks. Perhaps the best place to begin our story is with Aristotle, who lived in the fourth century BCE. He thought a lot about gravity, although not in a modern way. He believed that each effect must have a cause. His theory of gravity was tied intimately with the cosmological theory he devised. In this theory, the Earth was at the center of the universe. Objects fall, not because they felt a force in the modern sense, but because they had a natural place, which was at the center of the universe. This was true only for objects that were made of the element earth (of fire, water, air, and earth fame). Objects made of, for example, fire had a natural place in the heavens, which is why flames seem to be yearning for the sky. The ancient Greeks also believed the intuitively attractive, but ultimately false, idea that heavier objects fall faster than light ones.

While there were many individuals throughout the world who thought about gravity over the millennia, the next significant contributor to our understanding of gravity was born nearly two thousand years after Aristotle. This scientific legend is Galileo Galilei, an Italian genius born in Pisa Italy in 1564.

Galileo was born into a world and time of vibrant intellectual innovation—the Italian Renaissance. Learned men of the era explored many fields of science and mathematics, from astronomy and physics, to engineering and chemistry. While Galileo was influential in applying mathematics to physics problems, he also pioneered the use of the telescope in astronomy, setting in motion a successful scientific methodology that we still employ today.

But it is Galileo's investigation into the nature of gravity that is of concern to us, for he corrected a misconception that existed for two thousand years. He showed that heavy and light objects fell at the same rate.

Legend has it that he demonstrated this principle by dropping a heavy and light object off the famed Leaning Tower of Pisa, but this appears not to be true. No record of such an experiment is to be found in Galileo's writings, appearing only decades later in the writings of a former student. Instead, it would appear that Galileo's insights arose from him rolling objects down inclined planes. It's a pity that the legend involving the Leaning Tower isn't true, but that's history for you.

Galileo also postulated that objects falling in a vacuum would accelerate uniformly, and he was also responsible for some of the first claims that objects in motion tend to stay in motion. Indeed, his improvements in the understanding of the laws of motion paved the way for the next influential contributor to a modern theory of gravity—Isaac Newton.

Isaac Newton is arguably the most influential physicist of all time. Born about a year after Galileo's death, Newton revolutionized a huge swath of science, pioneering the laws of motion, inventing calculus, investigating optics, and—most importantly for us—working out the nature of gravity.

Before Newton made his contributions, there were many issues with how gravity was understood. To begin with, people didn't think that there was a single gravitational phenomenon. Instead, there were two. There was celestial gravity, which guides planets and comets through the heavens night after night. Learned men speculated that these objects were moved by the motion of angels' wings. And then there was terrestrial gravity, which governs the flight of an arrow and the path of tasty tidbits dropped from a toddler's highchair to a hungry pet sitting below. These two phenomena seemed to have nothing to do with one another.

However, Newton saw that the two were one and the same. Indeed, the very name of Newton's theory, for example, the unified theory of gravity, tells an important tale. A key facet of Newton's accomplishment is that it is the first example of a case where seemingly independent phenomena were shown to arise from a common origin. In short, among Newton's many, many, accomplishments, his unified theory of gravity can be considered one of the first steps toward devising a theory of everything.

Newton's theory is simple, at least in terms of the big picture. Two massive objects will feel an attraction between them. Increase the mass of either object, and the force of attraction will increase.

Bring those two objects closer to one another, and the force will also increase. Conversely, if you increase the distance between the two objects, the force will decrease.

Newton codified these observations into a mathematical formula that is still used today. We don't need to describe the equation here; rather, we merely need to understand the kinds of predictions that his work has been able to make. To begin with, we can calculate with excellent (although not perfect—more on that later) precision, the location of the planets at any time. We can predict the exact moment of solar eclipses and the location where they will occur, centuries in advance. We have sent men to the moon and have even fired a probe to the dwarf planet Pluto. That was a journey of about three billion miles, and when the probe arrived, it hit a target just a few miles wide.

Newton's equations can predict the velocity of stars orbiting the center of galaxies out to a distance of a few tens of thousands of light-years. Explaining the motion of stars at larger orbital radii is a little trickier, and we'll revisit that in a couple of chapters. This galactic success is certainly amazing. There's a reason that many would argue that Newton is physics' G.O.A.T. (greatest of all time).

While Newton's work is still the basis of essentially all modern engineering and much of physics, it turns out that his theory isn't quite perfect. First is a technical point, which is that it makes imperfect predictions in regions with strong gravitational fields. Second, it explains *how* gravity works, not *why*.

In Newton's theory of gravity, there is no explanation of how the force caused by the Sun is transmitted to the Earth (and vice versa). For things like sound, we explain the passage of sound waves as the bumping of (for example) air molecules into one another. However, in space, there is no air, no nothing. Newton called this "action at a distance" (*action* being the seventeenth century word for "force"). To resolve these minor difficulties, we need to discuss the insights of another candidate for physics' G.O.A.T., Albert Einstein.

Einstein was born in Germany in 1879, and he was deeply interested in physics. What fascinated him most were two things: the motion of objects moving at high speed and the motion of light. In 1905, as a result of his pondering these topics, he devised his special theory of relativity.

Much has been written about Einstein's special relativity. In it, different observers see clocks tick at different rates. Objects appear to have different lengths. Perhaps the most unsettling consequence of Einstein's theory is that the speed of light is the same for all observers.

In truth, that last statement isn't really a consequence of Einstein's theory. It's actually one of two of his postulates. The second is that two observers who are moving at constant speed with respect to one another are both perfectly within their rights to consider themselves stationary and to claim that the other observer is moving.

Einstein's theory of special relativity has been tested to exquisite precision. Special relativity makes predictions that are quite counterintuitive to most people and, even today, you can find on the internet an enormous amount written, claiming to disprove Einstein. Don't believe it. As weird as special relativity seems, it accurately predicts the outcomes of experiments.

However, Einstein wasn't satisfied with working out the laws of motion for objects moving at constant speeds. For our discussion on gravity, we need to talk about what happened when he turned his attention to what happens when objects are accelerated.

He began with thinking about what a person would experience if they were in an accelerating rocket with no windows. Such a person would feel a force toward the floor. And, if the rocket were accelerating exactly the right amount, the individual wouldn't be able to tell the difference between being in the accelerating rocket or standing in the same rocket, stationary on Earth.

You've probably never been in a rocket, but you can get a feel for the phenomena if you recall what happens when you get into a high-speed elevator. As it heads upward, it seems as if you weigh more than you do when you got in. That's because you are feeling both the gravity caused by the Earth and the force caused by the elevator's acceleration.

From these insights, Einstein was able to devise his theory of general relativity—a theory much like his special theory, but this time allowing for the two observers to be accelerated. He published this in 1915, a full decade after his publication on special relativity. This is a reminder that, even for geniuses, scientific progress can take some time.

The details and history of Einstein's work are fascinating, and there are valuable references in the Suggested Reading that delve more deeply into the technical and historical aspects. However, what we are most interested in is the punchline. What did Einstein's work have to do with gravity?

While Einstein is a household name, he wasn't the person who had the initial key insight. Hermann Minkowski, a mathematician born in what is now Lithuania, was a professor of Einstein's at the University of Zurich. He studied Einstein's 1905 papers and realized that they could be more easily understood as an exercise in geometry. In 1908, Minkowski published a paper that presented a paradigm-shifting observation. He showed Einstein's equations implied that space and time were really the same thing. While one observer could claim that an object was stationary in space and simply experienced the passage of time, another observer might claim that the object was both moving *and* experiencing time. Space and time were intertwined, and thus the idea of spacetime was born.

Minkowski's insights were very powerful and helped Einstein devise his general theory of relativity. The key equation in the theory can basically be written as:

$$(\text{mass and energy}) = (\text{constant}) \times (\text{geometry of space and time})$$

The implication of this was clear to physicists of the era. Matter and energy could bend space and this meant that familiar gravity is merely the curvature of spacetime. Figure 2.5 illustrates this point.

When we see a ball follow a curved path when we throw it, what is really happening is that the ball is following a straight path in curved spacetime. That is, admittedly, a hard concept to get your head around, but we already kind of understand this.

The Earth is a sphere, which means that it has a curved surface. If you were able to walk around the equator of the Earth, it would be reasonable to say that you walked in a giant circle, which is, of course, curved. But, from the point of view of the geometry of the sphere, you simply walked in a straight line on the curved surface. Basically, that's what general relativity says—what appears to be gravity to us is actually just an effect of mass curving spacetime.

Now that's a hard claim to swallow. Such an assertion must have some experimental verification. What unexpected predictions has Einstein's theory made? Well, there have been many validations of

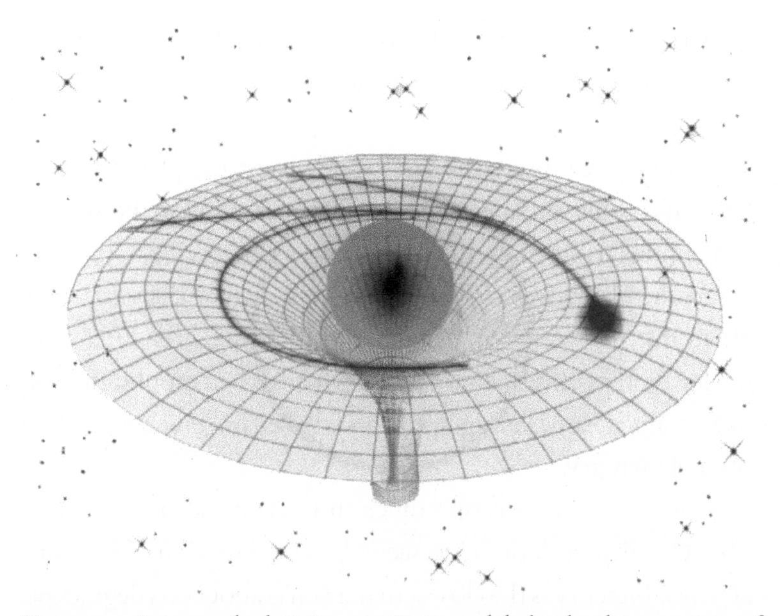

Figure 2.5 In general relativity, gravity is modeled to be the curvature of spacetime. Unlike Newtonian physics, in which planets orbit a star in a circular path, in Einsteinian physics, the planets follow straight paths in curved space.

general relativity—too many to mention in such a quick discussion. But let me tell you about just a few.

The first one was accomplished by Einstein himself. He turned this new theory to predictions of the orbit of Mercury. Naively, you'd think that Mercury just orbited the Sun in a circular path, but that's not right. Hundreds of years ago, in 1604, German astronomer Johannes Kepler realized that planets followed elliptical trajectories as they circled the Sun. And this is true of Mercury as well.

The mystery arose in 1859, when French astronomer Urbain Le Verrier analyzed a century and a half of observations of the orbit of Mercury, spanning 1697 to 1848. He realized that while Mercury was generally following an elliptical orbit, the orbit wasn't exactly elliptical.

In an elliptical orbit, the distance between the planet and the Sun is not constant. The planet is sometimes closer to the Sun and sometimes farther away. If the orbit were a perfect ellipse, the place where the planet was closest to the Sun would be the same, orbit after orbit. Le Verrier noticed that this isn't true: orbit after orbit, the location of closest approach slowly marched around the Sun, like a giant, cosmic spirograph. This is called the precession of the orbit of Mercury.

Now this precession isn't much. It is 532 arcseconds per century. One degree is 3,600 arcseconds, so the motion of the point at which Mercury is the closest to the Sun is about 0.15 degrees per century. That means it takes about 240,000 years for it to go once around the Sun. Among other things, this tells you that astronomical measurements were impressively accurate even centuries ago.

Astronomers calculated that the gravitational effect of Jupiter was responsible for most of Mercury's precession—specifically 489 arcseconds per century. But that leaves a discrepancy of 43 arcseconds per century that couldn't be accounted for. Astronomers of the nineteenth century hypothesized that the explanation was the existence of an undiscovered planet even closer to the Sun than Mercury. They even had a name for this hypothetical planet—Vulcan.

But when Einstein tried to calculate how much his theory of gravity affected the orbit of Mercury, he found that it was about 43 arcseconds, exactly the observed discrepancy. And, just like that, the evidence for the existence of Vulcan disappeared in an instant. Poof!

Predicting the orbit of Mercury is, without a doubt, an impressive feat, but it wasn't the one that launched Einstein into scientific superstardom. For that, we turn to the work of a British pacifist in a location far from home.

Sir Arthur Eddington (well, he wasn't a Sir at the time) was a gifted student. Born in 1882 into a Quaker family, his brilliance was recognized early on, and he received many honors during his undergraduate career. Because of his intellectual accomplishments, his career progressed rapidly. By 1913, at the young age of thirty-one, he was appointed as the director of the Cambridge Observatory, and he was made a member of the Royal Society the following year.

World War I began in the summer of 1914. Eddington spent the early years of the war doing astronomical research. Albert Einstein first described his theory of general relativity in 1915 and published it in detail in 1916. Because of the war, there was little communication between Germany and England, but Dutch physicist Willem de Sitter was in contact with both Einstein and Eddington, and it is through him that Eddington became aware of Einstein's work.

Eddington was fascinated by the theory and set about thinking of ways to test it. Because general relativity showed that gravity was the bending of space, Eddington realized that a strong enough gravitational field would distort the locations at which more distant stars were observed to be—conceptually no differently from how an ordinary lens works.

Because the strongest gravitational field around is created by the Sun, Eddington realized that this distortion effect would mean that stars that were located near the Sun in the sky would appear to be in different places than they would if the Sun wasn't there. Of course, one can't see stars during the day, so this idea was moot; that is, until Eddington thought about looking at the Sun during a solar eclipse. He could photograph the Sun during the eclipse and also capture nearby stars. He would then compare that photograph with ones taken six months later, when those same stars would now appear in the night sky. And, if the locations appeared to be different, that would mean that the Sun's gravity had affected the path of light from the distant stars to the Earth.

Eddington realized that an upcoming eclipse in 1919 would be a perfect test of Einstein's predictions. A total solar eclipse would begin in South America, cross the Southern Atlantic Ocean, and then pass on into Africa. He began to think about mounting an expedition to take the required photographs.

In March 1916, the British government began drafting men to join the war effort. Eddington tried to object to being conscripted on the basis of his pacifist Quaker beliefs, but this effort encountered resistance from the military authorities. Luckily, influential people in the scientific community intervened on his behalf and he was granted a deferment, as long as he organized and participated in the eclipse expedition.

The war ended in 1918, before the eclipse, but Eddington was committed. He hadn't agreed to make the measurement to avoid military service, but because he wanted to know the result. He boarded a ship that went to the island of Principe, off the west coast of Africa.

During the eclipse on May 29, 1919, Eddington photographed the Sun and the stars that were behind it. These were stars in the Hyades cluster, located in the constellation Taurus. Eddington returned to England to analyze his data and published it the next year. The data showed that the stars appeared to be in the "wrong" place because of the effects of the Sun's gravitational field. Figure 2.6 shows how it worked, while Figure 2.7 shows what Eddington saw.

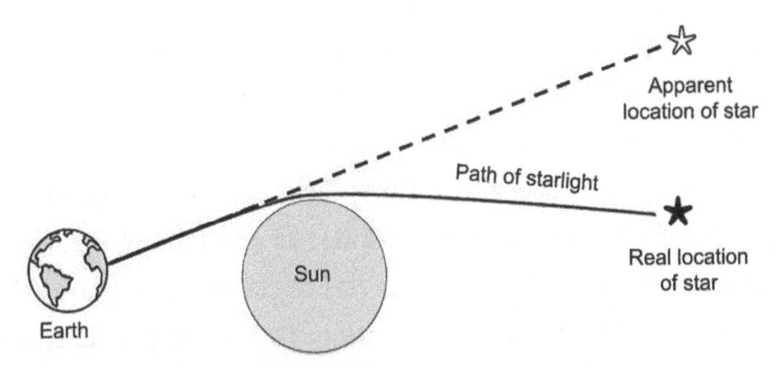

Figure 2.6 Eddington's experiment relied on the gravitational field of the Sun to make distant stars appear farther away from the Sun than they were.

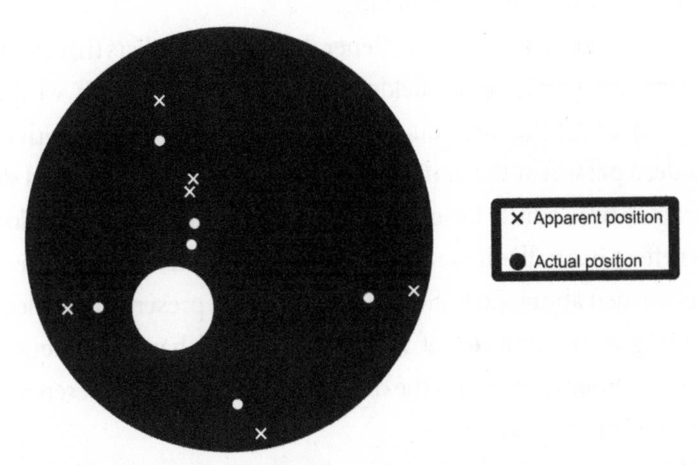

Figure 2.7 This figure shows what Eddington saw. The "x" symbols show the apparent location of the stars during the eclipse, while the filled circles show where the stars are located when not affected by the eclipse.

The result was published in the world's newspapers, including the *Times of London* on November 7 and in the *New York Times* on November 10, albeit on page 17 of the paper.

The media frenzy launched Einstein into the scientific stratosphere. He was well known in the scientific community, but regular people had no idea who he was. However, after the announcement of the Eddington experiment, one only needed to address a letter to "Albert Einstein, Germany," and it would appear in his mailbox.

While the effect of a gravitational field on the passage of light might have been one of the splashiest of the predictions of general relativity as far as the public was concerned, there are other predictions that are equally fascinating. You see, Einstein's theory also suggested that time was not experienced the same by all observers. His theory said that to do it right, you needed to take gravity into account. This is a trickier effect to understand, but one that is extraordinarily compelling.

On the face of it, it's simple. General relativity predicts that clocks in stronger gravitational fields tick slower than ones in weaker fields. Because the force due to gravity on Earth changes with altitude, a person at the altitude of the top of Mt. Everest would experience time a little faster than a person at sea level would. Now this effect is small—if we had super precise clocks since the Earth was formed about 4.6 billion years ago, in the present day, a clock orbiting at the altitude of Mt. Everest would have experienced about 39 hours more than the one at sea level. That's a difference of about one part in a trillion.

However, as tiny as that effect is, modern atomic clocks are precise enough to test this. There have been many such tests, including an iconic experiment in 1971 where two scientists put an atomic clock on an airplane and flew it literally around the globe. Upon their return, they compared the time recorded by the clock in the airplane to one on the ground and found that they disagreed exactly by the amount predicted by Einstein's theory.

My personal favorite test of general relativity is simpler than that. In 2010, scientists at the National Institute of Science and Technology in Colorado took two very precise atomic clocks, synchronized them, and then raised one of the clocks a single foot higher than the other. The raised clock experienced a gravitational field that was a tiny bit weaker, and it ran just a bit faster than the unmoved clock. This was a fantastic experiment, and it was yet another confirmation of a theory that was already believed by the scientific community.

If the justifications I've described aren't to your taste, maybe a practical example is of more interest to you. Perhaps you've used your phone to navigate as you've driven across the country. This navigation capability is made possible by the phone receiving radio signals from the global positioning system (GPS), which is a series of satellites orbiting the Earth. Each satellite contains an atomic clock. Your phone takes the clock signals from a couple of satellites and uses the time differences to determine where you are with a precision of about plus or minus two feet.

The satellites used in the GPS system orbit at an altitude of about 11,000 nautical miles, or about 20,000 kilometers, so they are in a region of weaker gravitational field. In addition, they orbit the Earth about twice a day, so they're moving very fast. Objects moving quickly are in the province of special relativity, and ones in different gravitational fields are in the realm of general relativity. Do these two theories play a role in the accuracy of the GPS system?

You bet. If GPS engineers didn't take into account the effect of relativity, the system would quickly become inaccurate. In a single day, the location GPS claims you were situated would be off by about six miles; on the second day, it would be twelve miles; and so on. So, to put it simply, if you are able to drive to grandma's house and not get lost, you should believe that Einstein's theory of gravity is real.

Another prediction of general relativity is, literally, Earth shaking. If matter and energy can distort space and time, what will happen if you have a hugely massive object moving rapidly? Say two small and compact stars that are gravitationally locked together in a very small orbit? The cyclical motion of those stars back and forth will cause rhythmic ripples in the fabric of space itself. These ripples are called gravitational waves, and they were first proposed by Albert Einstein in 1916, shortly after he developed general relativity.

Like any waves, gravitational waves carry energy; so, as two astronomical objects orbit one another, their orbital energy gets smaller as gravitational waves siphon it away. The reduced orbital energy means that the two objects get closer and closer to one another until they collide in a cosmic cataclysm. The gravitational waves generated by these collisions are enormous, and they spread out into the cosmos. Einstein reasoned that if humans built the right kind of detector, we'd be able to see the ripples in spacetime caused by these violent events.

The universe is big and the distances to other stars and galaxies are large. Like all waves, gravitational waves become weaker as they travel away from the source. On stellar or galactic distances, they become incredibly faint. In order to detect gravitational waves, they have to be either (1) generated close by, (2) incredibly violent, or (3) researchers need to build an incredibly sensitive detector. In the end, options 2 and 3 applied.

It is beyond the scope of this book to discuss in detail how gravitational wave detectors work. The interested reader should look to the Suggested Reading for more information. However, several highly sensitive gravitational wave detectors have been built, two in the United States, one in Italy, and ones in India and Japan are being constructed. The American research program is called LIGO (Laser Interferometer Gravitational Observatory), and the Italian one is called Virgo.

Thus, gravitational wave astronomers were ready on September 14, 2015, when cosmic gravitational waves passed through the Earth. The event was brief—just a fraction of a second. What caused this first observation of gravitational waves?

Literally a long time ago, in a galaxy far, far, away, two black holes were locked into a death spiral, orbiting and emitting gravitational waves that were too faint for researchers to observe. Then, after eons of dancing around one another, growing ever closer together, the two black holes merged, releasing an incredible amount of energy.

The two black holes were located about 1.3 billion light-years away. The black holes were massive—one containing 29 times the mass of the Sun, while the other contained 36 solar masses. In the final 0.2 seconds, the two black holes merged into a single black hole with a mass of 62 solar masses. The missing 3 solar masses was converted into the energy of gravitational waves. During that brief 0.2 seconds, the gravitational energy released was fifty times more than all of the light emitted by all of the stars and galaxies in the entire visible universe.

These gravitational waves spread across the cosmos at the speed of light, becoming ever fainter as they traveled. As they passed over the Earth, they caused tiny distortions in the shape of space—and I mean tiny. The four-kilometer-long LIGO detector stretched and shrunk by a distance smaller than 1/1,000 that of a proton. Yet that was within the sensitivity of the LIGO apparatus and gravitational waves were observed.

In the intervening years, many gravitational wave events have been observed. And the collisions weren't just between two black holes. Researchers have observed the merging of two neutron stars, as well as neutron stars and black holes. Gravitational wave astronomy is now an established discipline and, more importantly, their observation is yet another validation of the theory of general relativity.

There have been many, many more tests of general relativity. Within the accuracy of every test ever performed, the theory just works. Gravity is simply the bending of space and time. Weird, huh?

Let's take a step back and reflect. At about size scales ranging from the size of a human to the size of the cosmos, gravity reigns supreme. Einstein's theory can predict as different phenomena as the motion of a freefalling skydiver and the path of two nearby galaxies.

But accepting his theory means that we need to think about the cosmos and the nature of space and time differently. It's easy to think of space as something unaffected by the existence of matter. Our intuition tells us that space (in the geometrical sense) is where we exist, and the stars and planets simply pass through it.

But that's not what happens. Space and time are flexible, bending and twisting under the influence of matter and energy. Conversely, the geometry of space and time govern the path of astronomical objects. If you ask a cosmologist to concisely sum up Einstein's theory of gravity, they might say that spacetime tells matter how to move, while matter tells spacetime how to curve. And that's the story of gravity.

Now what?

So where are we in our quest to find a unified theory of everything? Well, we've actually made a lot of progress. The standard model of particle physics governs the subatomic realm and accurately describes the forces of electromagnetism and the strong and weak nuclear forces. The world of the small is a constant buzz of force-carrying particles jumping hither and yon, governing the behavior of denizens of the microrealm. Gravity is simply too weak to have much impact in the world of the small.

On a grander and more cosmic scale, it is gravity that reigns supreme. Huge and massive astronomical bodies like planets, stars, and galaxies move throughout the universe, warping the very fabric of space and time as they move. In this realm, the forces of electromagnetism and the two nuclear forces have little effect.

And that is kind of where we are. We have two very sophisticated theories that apply on different size scales—the tiny and the huge. Each theory works very well on its own scale and is nearly silent in the other.

It's possible that we could consider the job done. We can explain most of the behavior of the matter and energy we can observe. We've unified electricity, magnetism, chemistry, and light under the single theory of electromagnetism. We've unified electromagnetism and the weak nuclear force into the electroweak force. We've unified the behavior of all subatomic particles that are made of quarks as being governed by the strong nuclear force.

On the cosmic scale, we've unified celestial and terrestrial gravity into a unified theory of gravity. It is no longer necessary to invoke the beating of angels' wings to explain the path of a comet.

In short, we've made huge progress in our quest to invent a theory of everything. But we still have a handful of forces and a few building blocks. Perhaps we can do better and reduce the numbers of both? In our next chapter, we'll take a hard look at what many scientists hope is a path that will get us closer to the deepest truth.

CHAPTER 3

FAILED AND INCOMPLETE THEORIES

In the last chapter, I spent a lot of time doing what one might consider to be looking backward. Of course, that's not the direction we want to go, but by knowing where we've been, we often get a sense of where the path forward will take us. And, of course, the goal of fundamental physics is to learn more about the universe, culminating in that tantalizing theory of everything. Looking back can only get you so far.

It was American baseball player Yogi Berra who said that it's hard to make predictions, especially about the future. And it's worth remembering that science is an exploratory endeavor. While a researcher may have an idea as to what the outcome of a new experiment might be, there is always the exciting prospect that something entirely unexpected will be observed.

This has happened many times in the past. It was the case in 1895 when Wilhelm Roentgen noticed a faint glow on a screen coated with barium platinocyanide. The unexpected fluorescence led to the discovery of X-rays and the first Nobel Prize in Physics in 1901. And then there was the serendipitous cloudy day in March 1896, when French researcher Henri Becquerel was investigating how sunlight affected uranium salts. The overcast sky led him to put

Einstein's Unfinished Dream. Don Lincoln, Oxford University Press. © Oxford University Press 2023.
DOI: 10.1093/oso/9780197638033.003.0003

the salts and the film he was using to study them in a drawer. When he took the film out of the drawer a while later, he noticed that the film had been exposed. He had unwittingly discovered radioactivity. And there was the time in 1936 when Carl Anderson chanced upon the muon, opening up the modern world of particle physics. We heard that story back in Chapter 2.

In short, the history of physics—indeed, all of science—is replete with examples of unexpected discoveries. Thus, we might expect a surprise or two to arise as we push forward in our explorations, searching for that elusive, all-explaining theory.

But we have to start somewhere. What might a theory of everything look like, according to modern science?

Without a firm roadmap, scientists look at past successes in hopes that they will provide clues that might help future investigations. So let's quickly recap what we know and then think about how we might find out if the pattern we've seen in the past repeats itself. And let's first focus on the building blocks.

According to the standard model, the smallest building blocks known to physicists are the quarks and leptons. Quarks are found buried inside protons and neutrons, and the most familiar form of lepton is the electron. Most quarks and leptons are unstable and ephemeral and decay in much less than the blink of an eye. That's what we know. But are these particles the end of the road?

It is highly unlikely that in our search for an ultimate building block that quarks and leptons are the final word. Why do I say that? Well, every time we've found what we thought might be the final building block, we've eventually found out that there was a smaller building block inside of it. Is there any reason to think that this pattern continues with the quarks and leptons?

Well, yes and no. There is no direct evidence that quarks and leptons are composed of smaller objects still. Indeed, one of the first solid indications that they are built of smaller things would be finding that, rather than having the zero size that the standard model requires, quarks and leptons would have a small, but non-zero, size.

However, in spite of our best efforts, no such size has been found. Protons and neutrons have a radius of about a femtometer (10^{-15} meters). The Large Hadron Collider, or LHC—the most powerful particle accelerator on the planet—can resolve objects about 10,000 times smaller than a proton, or about 10^{-19} meters. The LHC has found no evidence that quarks and leptons have a size. Thus, quarks and leptons must be smaller than that.

And, of course, zero is smaller than 10^{-19}, which means the current measurements are completely consistent with the standard model. This means that they could have zero size. So is there any indirect evidence that quarks and leptons are composed of smaller things?

Again, the answer is yes and no. There are patterns in the properties of quarks and leptons that are suggestive that they have structure. I devote all of Chapter 7 to this particular topic, but I will sketch out the idea briefly here. When we look at the particles of the standard model, we note that there appear to be multiple copies of particles with similar properties, for example, the up, charm, and top quarks—all of which have a charge of +2/3. Similarly, there is the down, strange, and bottom quarks, with a charge of −1/3, and the electron, muon, and tau lepton, with their −1 charge.

Yet only the up and down quarks and electrons are commonly found in nature. The fact that there appear to be duplicates is highly reminiscent of the chemical periodic table of elements, in which we have chemically similar columns, like hydrogen, lithium, sodium, potassium, and so on. And we now know the reason for the similarities between the various families of chemical elements; it's because atoms have structure.

So what are the patterns seen in the particles of the standard model telling us? We simply don't know. Similar patterns in the chemical elements were also hinting at the full story, and it took over half a century—from Mendeleev's publication to the invention of quantum mechanics and nuclear physics—before it all made sense. We are in a similar place now with our understanding of quarks and leptons as our forebears were with their understanding of the chemical elements over a hundred years ago.

Many physicists think that perhaps these patterns will be resolved with the discovery of quark and lepton constituents, although this is by no means a universal belief. There are some who argue that quarks and leptons are the end of the line. I will talk much more about this in Chapter 7. However, the simple and honest truth is that we currently don't know what is going on. We simply see these patterns and ponder.

What about the Forces?

In Chapter 2, I told you of a number of instances where seemingly unrelated phenomena were shown to originate from a common cause. Newton unified celestial and terrestrial gravity. Maxwell unified electricity, magnetism, chemistry, and light. And a group of researchers in the 1960s unified electromagnetism and the weak nuclear force.

We're left with three forces: electroweak, the strong nuclear force, and gravity, with the open question hanging around of what drives the Higgs field. So maybe we might call it four forces. And, of course, many writers don't fully acknowledge electroweak unification, so some will split it up into its component electromagnetism and the weak nuclear force. So what future unifications do some physicists envision?

Well, the obvious answer is that we don't know what the final answer will be. But there has been much speculation over the decades. One oft-quoted idea is that we will find that the electroweak and strong nuclear forces originate from a single combined grand unified force, described by a grand unified theory, or GUT. Nobody knows what form the grand unified theory will take, so this idea is currently nothing more than interesting, wholly hypothetical speculation. This unification is thought to occur at an energy far higher than is accessible by even the most powerful modern accelerators.

I said that nobody knows what form a grand unified theory will take, and that's true. However, past researchers spent a lot of time trying to create such a unifying theory. The heyday of some of these theories was back in the 1970s and early 1980s. None of those early theories proved to be an accurate representation of reality, so they are forgotten today. A little later, I'll give a brief historical review.

A successful grand unified theory will still leave gravity standing alone. In order for a final theory to combine all of the forces, it will have to eventually address head-on the differences between current gravitational theory and the quantum realm. If successful, physicists will perhaps figure out how to unify the hypothetical grand unified theory and gravity. The result will be a theory of everything, or TOE. A theory of everything is the ultimate goal of fundamental physics.

Figure 3.1 illustrates the big ideas, showing currently accepted unifications in black lines and imagined unifications using white ones. And I cannot emphasize enough that these imagined unifications are just that—imagined. You should consider them to be quite reasonable, but far from proven.

Aside from the unproven idea of unification, it is perhaps of interest to spend a little time talking about the energy scales at which the unification might occur. And, to do that, we need to define an unfamiliar unit.

If you ever took a physics class, you learned about a unit of energy called the joule. A joule is a relatively small amount of energy on human standards. It takes 4.2 joules of energy to raise

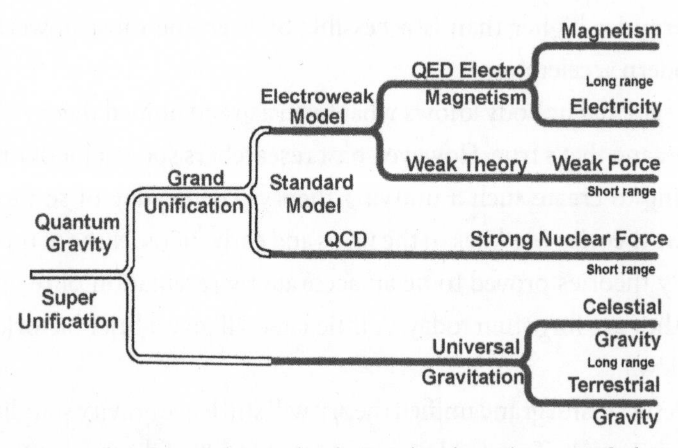

Figure 3.1 This figure shows the forces that have been unified so far (denoted with black lines) and those that the theoretical physics community expects will be unified in the future (white lines). The white lines have not been confirmed experimentally and should be regarded as informed speculation. QCD is short for quantum chromodynamics, and QED is short for quantum electrodynamics.

the temperature of one gram of water by one degree Celsius. (For Americans whose metric system is a little rusty, a gram is about the weight of a paperclip and a degree Celsius is just shy of two degrees Fahrenheit.)

While on human scales a joule is relatively small, it's an absolutely huge amount of energy on subatomic scales—and it's unwieldy to use. It's like using astronomical units when measuring the size of your waist. An astronomical unit is the distance between the Earth and the Sun, so you can see that it's a bad unit for a tailor to use as she is custom fitting your pants. So we need another, and more appropriate, unit to talk about subatomic interactions.

The unit of choice is the electron volt. An electron volt is the amount of energy an electron will gain if it is accelerated by an electric field with an electric potential of one volt. To give some sense of scale, to break apart a water molecule, you need to add about ten electron volts of energy. So we see that the unit is a reasonable one to work in the quantum world. Two additional examples of the energy used in subatomic processes are X-rays, which typically have an energy of a few tens of thousands of electron volts, and radioactive decay of atomic nuclei, which usually involves about a million electron volts. A joule is 6.2×10^{18} electron volts, so we see that it is really an inconvenient unit for a particle physicist to use.

The electron volt might seem to be a curious choice—after all, why has it been chosen, aside from being a less cumbersome unit for discussing quantum measurements? It has to do with particle accelerators.

Remember that physicists use particle accelerators to shoot subatomic particles at targets. All particle accelerators use electric fields to force particles with electric charge (like protons and electrons) to move. All electric fields can be described by their electrical voltage. If a one volt electric field can accelerate an electron or proton to an energy of one electron volt, a hundred volt field will result in a hundred electron volt particle. A million-volt electric field will give a subatomic particle a million electron volts of energy. So we see that in addition to the fact that the electron volt is a more natural unit for discussing subatomic physics, it's also easier for physicists who want to know how much energy their particle beam has. If you know the voltage of the electric field of your accelerator, you know the energy of your beam.

Modern particle accelerators can accelerate particles to very high energy. The details of how they do that are a bit more complicated than the simple description I've given above, but engineering details are not central to the topic of this book, so I'll forgo describing the technology here. If you're curious, there are some references in the Suggested Reading.

The most powerful particle accelerator currently operating can accelerate beams of protons to nearly ten trillion electron volts of energy. Given that that's a mouthful, physicists truncate that phrase as TeV, or teraelectronvolts. ("Tera" being the metric prefix for "trillion." When physicists talk, they pronounce each letter, e.g., T-E-V.)

If 10 TeV is approximately the operating point of the highest energy particle accelerator ever made, how does that compare with the energy that scientists think that the grand unified theory and theory of everything become important? While this is entirely speculative, current thinking is that the grand unified theory scale is about 10^{12} TeV, and the theory of everything scale is about 1.2×10^{16} TeV.

These energy scales are much higher than we can currently create in our laboratories. That last one, the energy scale at which it is commonly thought that a theory of everything will apply has a special name. It's called the Planck energy. So just what is the Planck energy, and why is it that many scientists think that it's the right energy to start talking about a theory of everything?

Planck Units

The story of Planck units is an interesting one, which should be told in two parts. The second half is rarely shared, but it is key to understanding the connection between Planck units and a theory of everything. This connection is one that the creator of the units could never have imagined.

In 1899, German physics legend Max Planck turned his attention to a mystery that was considered to be very pressing at the time. His work set scientists on a path that resulted in the field of quantum mechanics. So what was that mystery?

The mystery involved, of all things, the color of light emitted by kilns. High-temperature kilns are essentially ovens that are used to convert dried clay into crockery. These ovens are not your ordinary kitchen oven. Some of them are heated to such a high temperature that the inside of the oven glows.

The prevalent physics theory of the late nineteenth century predicted that the color of light emitted by these ultra-hot ovens should be blueish or purple; however, measurements very clearly showed that there was a lot less blue light found in the ovens' glow than expected.

Planck devised a new theory, which agreed with measurements. Unlike the previous theory, which postulated that all colors of the rainbow carried equal energy, his theory said that short-wavelength light (on the blue side of the rainbow) carried more energy than long-wavelength light (on the red side). The amount of energy carried by each color is a combination of how bright the color is (i.e., how much of it is there) and how much energy that color carries. If a wavelength has more energy, you need less of it to account for the energy carried by that color and therefore that color is dimmer. And, from Planck's humble postulate, quantum mechanics was born.

Planck's theory included an equation that related the energy of a particular color of light to its frequency or, using the appropriate relationship, its wavelength. This equation was simple and just said that energy was equal to a constant times the light's frequency. That constant is now called the Planck constant. While Planck treated it as a constant of proportionality, we now know it has broader implications and is a useful measure of subatomic angular moment or what we colloquially call spin. The symbol for Planck's constant is h and the spin of, for example, a photon is simply $h/(2\pi)$. In modern times, we have defined the so-called reduced Planck's constant as $\hbar = h/(2\pi)$. This symbol is simply called "h-bar."

While the creation of quantum mechanics is a fascinating story, it is a mere sidebar in the discussion of the Planck units. You see, there was another topic that interested Planck and that was the question of physical units like length, mass, time, and so on.

Historically, units have been a bit arbitrary. A foot was literally the length of an adult male's foot. A meter was first defined to be one ten-millionth of the distance from the North Pole to the equator, along a line drawn through the city of Paris. For units of mass or weight, there is the pound, which dates back at least to Roman times, and the kilogram, which was first simply defined to be the mass of a liter of water.

The arbitrariness of physical units troubled Planck, and he set out to figure out a set of units that would be the same for everyone and that were motivated by science. His approach was to take physical constants of the universe and use them to derive universal units.

He began with four physical constants, specifically: G, which is the gravitational constant and is found in Newton's equation for gravity; c, which is the speed of light; $\hbar$, the reduced Planck constant; and k_B, the Boltzman constant, which appears in the study of the energy stored in gases.

Planck took these four physical constants and multiplied and divided and took square roots until he could come up with combinations that had the units of length, time, mass, and temperature. These are shown in Table 3.1, along with the Planck energy.

Table 3.1 This table shows the manner in which one can take ratios of the gravitational constant (G) the speed of light (c), the reduced Planck constant ($\hbar$), and the Boltzmann constant (k_B) and extract some of the Planck units.

Name	Expression	Value (Metric Units)
Planck length	$L_P = \sqrt{\frac{\hbar G}{c^3}}$	1.6×10^{-35} meters
Planck energy	$E_P = \sqrt{\frac{\hbar c^5}{G}}$	2×10^9 joules (1.2×10^{16} TeV)
Planck mass	$M_P = \sqrt{\frac{\hbar c}{G}}$	2.2×10^{-8} kilograms
Planck time	$T_P = \sqrt{\frac{\hbar G}{c^5}}$	5.4×10^{-44} seconds
Planck temperature	$\Theta_P = \sqrt{\frac{\hbar c^5}{G k_B^2}}$	1.4×10^{32} kelvin

The nice thing about the Planck units is that they are universal. In fact, in his paper where he proposed them, Planck wrote "... it is possible to set up units for length, mass, time and temperature, which are independent of special bodies or substances, necessarily retaining their meaning for all times and for all civilizations, including extraterrestrial and non-human ones, which can be called 'natural units of measure.'"

Now that doesn't mean that all civilizations will come up with the same numerical value for any particular Planck unit. For the Planck energy, a non-American civilian will quote the 2×10^9 joule number. A particle physicist will quote 1.2×10^{16} TeV, while a Martian scientist might say that it is 5.9 zlorbs. However, since everyone will be calculating the same physical length in their own favored units, it is therefore easy to translate between different unit schemes. For instance, if the Planck energy is both 2×10^9 joules and 1.2×10^{16} TeV, then one can convert between the two by using the conversion factor of 6×10^6 TeV per joule.

While the creation of a universal set of units—what Planck called "natural units," is a very helpful achievement, it doesn't tell us why the Planck energy has anything to do with the scale at which a theory of everything might apply. For that, we need to turn to a paper that was published in 1964.

The year was 1959 and American Alden Mead was a chemist at the University of Minnesota when he asked himself how gravity would affect the operation of an incredibly precise microscope.

Ordinarily, gravity is of no consequence when using a microscope. What matters is how the wave nature of light spreads out when it passes through a small aperture. For the cognoscenti, this is called diffraction.

Mead employed quantum theory in his derivation. He knew that the Heisenberg uncertainty principle stated that you can't simultaneously know the position and motion of any subatomic particle. This is relevant when looking at a small thing. If you can resolve the small thing by shining a light on it, you have little information on the motion of the light. That's how Heisenberg's principle works—the better you know the position, the less you know about the motion, and vice versa.

When one includes gravity, the situation is a bit more complex. He considered what happens when you bring a photon very close to some subatomic phenomenon you're trying to image—say a super tiny particle, much smaller than anything we know of today. As the photon gets closer and closer to the object being imaged, the gravitational force between the photon and the object grows. Furthermore, because of the uncertainty principle, as the photon nears the imaged particle (and thus the location becomes more precise), the motion of the imaged object becomes more and more uncertain. Because its motion is uncertain, the details of how it interacts with the photon become more uncertain. The whole theoretical edifice of gravity and the uncertainty principle becomes untenable when one approaches the Planck length. When all effects are combined, the result is that it is impossible to image things smaller than the Planck length.

When Mead tried to publish his calculation, he encountered considerable resistance. Referee after referee would object to the importance of his result. He spent five long years trying to convince the scientific community and journal editors that not only had he done his calculation properly, but that it was of significant importance. His paper was eventually published in America's flagship physics journal, *Physical Review Letters.*

However, the significance of his result lay dormant for many years. In 2001, he wrote in a letter to the editor of the journal *Physics Today* of his experience. He was responding to an earlier paper by American physicist Frank Wilczek, in which Wilczek discussed the importance of the Planck length in theoretical physics. According to Mead's telling, in the 1960s, many physicists thought that there was no limit on how small things could be resolved, while some others thought that other phenomena defined the smallest things that could be seen. I spoke to Mead shortly after his letter to the editor was written, and he emphasized how difficult it was to get the community to accept his result.

And yet, in the intervening half-century, the community has embraced his paper. The current thinking is the following. First and foremost, Mead's result sets a lower limit on the smallest size for which our current theory of gravity can apply. The one irrefutable conclusion is that if no new physical phenomena are discovered, general relativity cannot apply for sizes smaller than the Planck length. Modern physics theories completely fail for that size and smaller. This is generally accepted.

However, many in the theoretical community have taken that conclusion one step farther. They believe that since the Planck energy scale and Planck length is the point at which current gravitational theory fails, it is at those scales that a theory of everything will become evident. I think that this conclusion, while defensible, is a bit on the hasty side, as we will soon see.

I should like to say that I have glossed over most of Mead's argument in the interests of brevity and focused merely on the big ideas. For the person who wants to dig much deeper into his work, I have included Mead's original paper and his interchange with Wilczek in the Suggested Reading.

Moving On: Early History

At this point, we've finally gathered enough knowledge to start digging more directly into the quest for a theory of everything. I'll start with some early thinking and explorations but will then move to more modern efforts.

So let's start with a fundamental question. If we have a successful quantum theory of the electroweak and strong forces, and a successful non-quantum theory of gravity, why should we expect that we even need a quantum theory of gravity? Maybe they're just very different—gravity deals with the nature of space and time, while the other forces exist inside that spacetime. It's not at all obvious that quantum gravity must even be a thing.

Now in the previous section, we determined that at the Planck energy and length our current theory is irretrievably broken, but that doesn't mean that whatever replaces it must be quantum in nature. It could simply be some different theory without quantum properties. Surely there must be some reason why physicists talk confidently about the need for a theory of quantum gravity.

Perhaps the simplest argument for quantum gravity is that all of the other forces have a quantum nature. After all, when electromagnetism is first taught to introductory students, there is nary a hint of quantum behavior—that's the same for gravity. But scientists saw behaviors in the emission of light that classical electromagnetism couldn't explain—the previously mentioned color of glowing kilns, for example. Are there lessons that the development of ordinary quantum mechanics can teach us?

Let's start out with one of the consequences of a common (and basically classical) vision of the atom. It is not at all uncommon for people to imagine an atom as an atomic nucleus being orbited by electrons, what is called the "Saturn" model of the atom. This model was proposed in 1904 by Japanese physicist Hantaro Nagaoka. The electron had been discovered only a few years prior in 1897, and the discovery of the atomic nucleus lay in the future. (Ernest Rutherford discovered it in 1911.)

The Saturn model of the atom was short lived, as classical electromagnetism showed that it was unworkable. According to Newton's first law of motion, an object in motion will continue to move in a straight line unless it is acted upon by an outside force. That force will cause the object to accelerate.

An electron orbiting a nucleus is obviously not moving in a straight line. The electric force between them would cause the electron to move in a curved and, most importantly, an accelerating path.

Classical electromagnetism—as in the electromagnetism of Maxwell discussed in the previous chapter—showed that an accelerating electrical charge will emit radiation and thereby lose energy. Countless experiments have validated this prediction.

But that's a problem for the Saturn atom. If an electron is orbiting a nucleus, the constant acceleration, radiation, and loss of energy will cause the electron to spiral down into the nucleus in about twenty-trillionths of a second.

But atoms are stable. Among answering many other questions, quantum mechanics was devised to come up with a theory that didn't predict that atoms would self-destruct. In short, quantum mechanics saves atoms from theoretical oblivion.

For gravity, there is a similar problem. According to general relativity, an accelerating mass will emit gravitational radiation. And, if electrons move in the vicinity of a nucleus, then they will, by virtue of their acceleration, also emit gravitational radiation. This requires a solution. This solution doesn't have to be quantized in nature, but there must be a better theory that explains why gravity doesn't destroy atoms.

It should be noted that ordinary quantum mechanics doesn't save us here. Ordinary quantum mechanics (really the entire standard model) only safeguards the known quantum forces. It is silent on gravity. So either we need to modify our theory of gravity, or there needs to be some deeper connection between gravity and the other known forces so ordinary quantum mechanics saves both. If so, that would be a clear step toward a theory of everything.

There's another reason that we need to devise a subatomic theory of gravity and that's for another simple reason: general relativity totally fails in the microcosm. As particles get closer and closer to one another, the theory predicts that their force becomes larger and larger, edging toward infinity. In fact, there have been a number of attempts to try to modify gravitational theory for a quantum environment, and all of the theories generate infinities in the equations. A few generate an infinite number of infinities, which is, as my grandmother used to say, quite a pickle.

I hope that I've convinced you that general relativity fails in the subatomic world. I probably haven't convinced you that a theory of microscopic gravity must be quantized in the sense that the other forces have, but it's likely that an equivalent approach will apply for gravity. Or maybe not. We won't know until we've succeeded.

On the other hand, we've certainly not succeeded so far. On the third hand, smart people have tried. Let's spend some time getting acquainted with a sampling of earlier efforts to construct theories of quantum gravity, grand unified theories, and even theories of everything.

Einstein Fails

Albert Einstein is arguably one of the greatest physicists of all times and certainly of the twentieth century. The year 1905 was his *Annus Mirabilis*, his miracle year. In that year, he received his PhD and wrote four influential scientific papers. He postulated that light consisted of individual particles that we now call photons. It is for that work that he received his Nobel Prize in Physics in 1921. He wrote a paper about what is called "Brownian motion," which is the motion of dust motes seen when viewed through a microscope. His paper explained Brownian motion as the bombardment of the motes by atoms, the existence of which had not been definitively established at that time.

However, it is his other two papers written that year for which he is more famous. They were his special theory of relativity and the related concept that mass and energy were actually the same. He generalized this work a decade later, when he published his general theory of relativity. We learned a bit about this in the previous chapter.

While the first two decades of the 1900s were Einstein's heyday, the 1920s were the era of quantum mechanics. And Einstein really didn't like quantum mechanics. Unlike classical physics, quantum mechanics said that the universe was inherently probabilistic. While Einstein was not alone in his distaste for the idea, he certainly was one of the more stubborn and vocal of the critics.

Now, it's not that Einstein rejected the mathematics of quantum mechanics. After all, he could see as well as anyone that they accurately predicted the measurements that his experimental colleagues were making. However, he regarded quantum theory as incomplete and that a more sophisticated and comprehensive theory would better explain the laws of the universe and would generate the disconcerting quantum theory as a consequence.

Armed with the firm conviction that there exists a single unified theory that would explain all other theories, Einstein embarked on a thirty-year journey trying to find it. When he began his investigations, only two of the four forces we know today were known: electromagnetism and gravity. And Einstein was the undisputed king of gravity.

What he tried to do was to unify electromagnetism and gravity. While the majority of the physics world was enthralled with the developing field of quantum mechanics, Einstein and a few colleagues worked to unify gravity and electromagnetism.

In 1919, German mathematician Theodor Kaluza was niggling around with Einstein's theory of general relativity and, by niggling, I mean exploring it in sophisticated ways.

By that time, Einstein had completely adopted Minkowski's mathematical formulation of his theory. General relativity was most easily understood as the intricate interplay between matter and energy and a four-dimensional spacetime. Four dimensions—that was key.

But Kaluza was a mathematician, so he decided to see what would happen if he expressed Einstein's theory in not four, but rather five, dimensions. He was shocked to find that if he did so, not only did general relativity fall out of his calculations, so did Maxwell's equations of electromagnetism. He communicated his accomplishment to Einstein, who encouraged him to publish it. Kaluza's advance, while exciting, had a rather serious flaw. We don't live in a universe that has five dimensions of spacetime.

Swedish theoretical physicist Oskar Klein became aware of Kaluza's work and also its fatal flaw. Klein, working at the interstices of physics and mathematics, had a clever idea. What if Kaluza was correct, and there are five dimensions of spacetime, but not all dimensions were the same? Suppose that three of Kaluza's dimensions were ordinary space of infinite extent, one was time—also infinite, but the fifth dimension was curled up and thus very small. In a two-dimensional analogy, he imagined spacetime was like a tightrope at a circus, with one direction along the rope very long, but the other (around the circumference of the rope) very short. The resulting combined theory is now called Kaluza-Klein theory and it was forgotten for decades; however, some of the key ideas were resurrected as superstring theory was developed. Superstring theory is a modern candidate for a theory of everything and we'll return to it later.

Einstein liked Kaluza-Klein theory very much and returned to it frequently over the years, but he never could find a way to extract quantum mechanics from it. So he explored other options as well. Another approach in which he invested a great deal of time retained his four-dimensional spacetime but tried to generalize his relativistic equations to include electromagnetism, without success.

Einstein spent over three decades trying to uncover a theory of everything, trying many ideas and having little success with any of them. "Most of my intellectual offspring end up very young in the graveyard of disappointed hopes," he wrote in a letter in 1938.

It might appear unseemly for an ordinary physicist to criticize the legendary Einstein, but his approach was doomed to failure. You see, by the 1920s, physicists knew of more than just two forces. The existence of protons inside the nucleus of atoms implied the existence of the strong nuclear force, and beta radiation similarly implied the existence of the weak nuclear force. And there was also that pesky field of quantum mechanics that the majority of atomic physicists were developing.

Einstein completely ignored the nuclear forces and, while he acknowledged quantum mechanics, he dismissed it as being nothing more than a curious consequence of the deeper theory which he pursued. In short, Einstein spent the last thirty years of his life walking a path that we now know to be unproductive. In contrast, the bulk of the physics community embraced these other phenomena and made great progress in our understanding of the laws of nature.

Einstein increasingly became a curiosity, acknowledged as brilliant, but no longer in the center of things. He spent the rest of his life trying to unify gravity and electromagnetism, even working on it on his deathbed. He died in April 1955 in the hospital from an abdominal aortic aneurysm, leaving behind an unfinished dream.

1970s Grand Unified Theory Ideas

The search for unification didn't end with Einstein's death. Other ideas have been tried.

The 1970s and 1980s were the decades of bell bottoms and big hair, but they were also the heyday of a new approach to unification. The physicists of the era weren't reaching for the big ring, so to speak. They weren't trying to develop a theory of everything, but rather a grand unified theory, which, it was hoped, would bring together the strong force and the electroweak force. The key insights and techniques used in this effort were highly mathematical and abstract and go by the names symmetries and group theory.

Because these ideas are so inherently mathematical, I will employ a generous dollop of analogies to convey the key points. Just realize that they are analogies and only apply so far. I will focus predominantly on the symmetry side and mention the group theory aspect only so you are familiar with the terminology.

Symmetry has both a mathematical and aesthetic meaning, and the two meanings are similar. The right and left sides of a specific human body are essentially identical. If you look at a person head-on or if you look at them in the reflection of a mirror, it's hard to tell which is which. We say that the human body is symmetrical to left/right flips.

Here's another analogy that demonstrates both symmetry and group theory. Suppose you're in a room with no windows and you have nothing with you except your clothes. On the floor are two black lines of identical length that cross one another to make the letter "X." The two lines are exactly perpendicular to one another.

If I asked you to pick one of the "arms" of the X and label it as "north," it wouldn't matter which of the arms you picked—they're all the same. That's symmetry. The crux of symmetry is if you make a change, you don't notice. Note that the direction you chose has nothing to do with earthly north . . . it's just a reference frame for inside the room. Perhaps it should be called "room north," but we'll just stick with "north" to keep it simple.

Now, when you do that, you know that to face east in the traditional way, you'd have to turn to your right by 90°. South is 180°, and west is 270°. Group theory tells you what happens if you turn right or left. Turn right 360°, and you're back looking north again. Turn right 540°, and you're looking south. Or turn right 90° and then left 180°, and you're now facing west. Group theory tells you about how combinations of right and left turns work and what the final direction is.

Symmetries are found throughout physics. Sometimes we understand what they are doing and why. Sometimes we don't. In those cases, observing symmetries might give hints to theoretical physicists on how they might develop a new theory. And sometimes the innovation goes the other way, with researchers discovering symmetry in an existing theory.

One example of the second path occurred in 1932, when German physicist Werner Heisenberg observed that neutrons and protons seemed to experience essentially the same amount of strong nuclear force when they were placed in a nucleus. (In this context, when I say the strong nuclear force, I mean the force that holds together protons and neutrons, not one that acts on quarks. The two forces are related, but different. Unfortunately, different books use the term differently and that can be confusing.)

If the proton and neutron experience the same amount of strong nuclear force, then in the theory for that force, they can be swapped without anyone realizing the difference. (It's perhaps important to note that the proton and neutron can be distinguished, as they have slightly different masses and experience the electromagnetic force differently. Thus, Heisenberg's insight applies *solely* to the two particles' behavior under the influence of the strong nuclear force.)

When one accepts that the proton and neutron are interchangeable under these conditions, we can replace their names with the generic word "nucleon," which just means "particle in the nucleus" and call the proton to be an up-type nucleon and the neutron to be a down-type. Up and down have no real meaning here, other than "opposite."

If the two particles are interchangeable, then it is inevitable that the theory will predict that there exists some method for them to morph into one another.

In 1935, Japanese physicist Hideki Yukawa devised a theory that predicted this morphing. He proposed that the strong nuclear force was carried by three particles that we now call "pions" (short for pi mesons). The three different pions had different electrical charge: positive, negative, and neutral.

A proton (which has positive charge) could emit a positive pion and thereby turn into a neutron. A neutron could absorb that positive pion and turn into a proton. And if a proton emitted a neutral pion, it remained a proton. These transitions are illustrated in Figure 3.2. This changing back and forth has been observed, and it will become important when we get to the grand unified theory efforts of the 1970s.

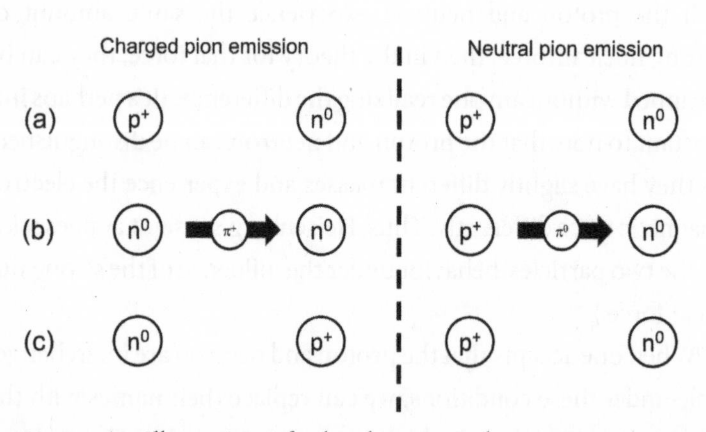

Figure 3.2 An illustration of what happens when a proton emits a charged pion (left) or neutral pion (right). (a) is prior to the emission, (b) is during the emission, and (c) is after the pion is absorbed. The exchange of a charged pion changes both the emitter and receptor's identities, while the exchange of a neutral pion changes neither.

Pions were not discovered until 1947 (charged pions) and 1950 (neutral pions). They were just one of the many particles discovered in cosmic ray and accelerator experiments during the 1950s and early 1960s. However, the most important point is to remember the connection between symmetries and theories. The nucleon symmetry proposed in 1932 led to the first successful nuclear theory proposed in 1935.

Heisenberg and Yukawa's contributions to nuclear theory have both a symmetry and group theory nature, but they are not the most striking examples. For that, we need to turn to the middle part of the twentieth century, when dozens of new particles were being discovered.

Initially, it would appear to be a lost cause to try to find order among the chaos. There were particles with charges spanning the range of −2, −1, 0, +1, +2 times the proton. Masses ranged widely—from the lowly electron, with a mass of 1/2,000 that of a proton, to a few with the mass of about 1/10 that of a proton, and then even more ranging up to just shy of double that of a proton. The particles had subatomic spin that were integer or half-integer multiples of the reduced Planck constant $\hbar$. Some particles would decay in the breathtakingly short time of 10^{-23} seconds, while others lived for the "long" time of about 10^{-6} seconds. Then there were the stable proton and electron. And, of course, some of the particles were "strange," meaning that they were easy to make, but they decayed very slowly. Some of them were "doubly strange," which meant that they were strange, but so were at least one of their decay products. We now know that this strange property is because the particle contained strange quarks, but they didn't know that back in the 1950s. It really was a mess.

So what was a puzzled physicist to do? They took the dozens of particles known at the time and tried to arrange them according to patterns. There were many, but let's take one as an example. These are particles with a subatomic spin of 1/2, in the mass range near that of a proton. The electric charge and strangeness of the particles were not constrained. All of the particles were of the class called baryons, which simply means they were heavy. There were the familiar proton (p) and neutron (n), four particles called the sigma (Σ) and lambda (Λ) baryons. All four of these were strange particles. Finally, there were doubly strange "cascade particles (Ξ)." They were called cascade particles, because of the fact that both they and their daughter particles were strange.

If you take these eight particles and arrange them by both the amount of strangeness (S) they have, and their electrical charge (Q), they form a hexagon, as illustrated in Figure 3.3.

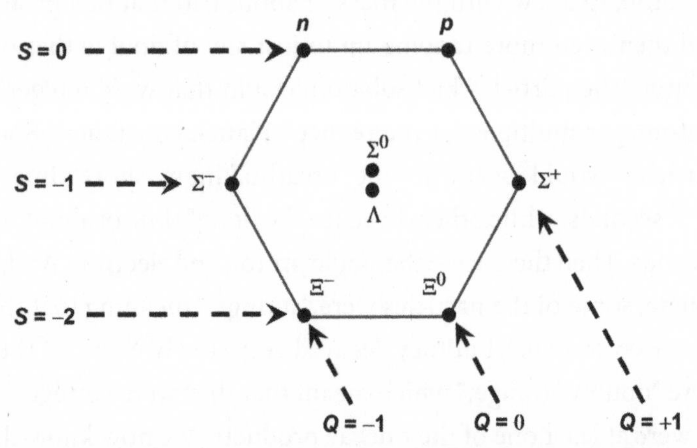

Figure 3.3 Baryons with spin ½, arranged by charge (Q) and strangeness (S). We see the familiar neutron (n) and proton (p), as well as six other, less familiar baryons. The structure seen here clearly suggests that there are some rules that govern which baryons can and cannot exist.

There isn't anything physical about the hexagon. It's just an organizing principle. However, you can't look at that figure and not wonder just exactly what message it's telling you. And that's not the only geometrical figure that arises when scientists scrutinized other particles. Baryons with spin of (3/2) $\hbar$ made a similar diagram that was a triangle. And other geometrical figures arose when the process was repeated with particles with subatomic spin that was an integer multiple of $\hbar$. The plethora of geometric structures simply screams, "There are underlying principles that govern these particles," much like the chemical periodic table hinted at the atomic structure we know today.

As we discussed in Chapter 2, these patterns were explained in 1964 when Murray Gell-Mann and George Zweig proposed the first three of what Gell-Mann called quarks. There was the up (u), down (d), and strange (s). And the particles shown in Figure 3.3 can be explained as being different combinations of three of those quarks. Table 3.2 shows the quark content of these eight particles.

The first version of quark theory was simple. It treated the three quarks identically. If they were identical, then it would be possible to change the quarks in a specific particle and the new particle would be identical. That would mean that a proton, with its *uud* content, could turn into a neutron, with its *udd* content, by changing an up to a down quark.

If this were strictly true, the group theory mathematics that governed the three original quarks would be called SU(3). SU(3), short for "special unitary group, with three elements." That comes with a lot of mathematical baggage, which is well beyond the scope of this book. Suffice it to say that SU(3) is a compact way to

Table 3.2 The quark content of the eight baryons shown in Figure 3.1.

p (*uud*)	n (*udd*)	Ξ^- (*dss*)	Ξ^0 (*uss*)
Λ (*uds*)	Σ^-(*dds*)	Σ^0(*uds*)	Σ^+(*uus*)

say "a group of three elements that can be swapped without anybody noticing."

Of course, in the case of the original quark theory, SU(3) was only an approximation and one that was clearly imperfect. After all, the quarks have different electrical charges and the strange quark is heavier than the other two, which is why a proton is different from a neutron and why the strange baryons in Figure 3.3 are heavier than the familiar nucleons.

Maybe an analogy will help. SU(3) models reasonably well the ways in which you can assemble quarks, but imperfectly. It's like in the United States, where all citizens are nominally accorded equal rights, but socioeconomic differences mean that we do not perfectly achieve the ideal. Similarly, with the three quarks known in the early 1970s, there was a symmetry, but it was imperfect.

And, of course, we now know that there exist at least six quarks, and the six quarks have wildly different masses. Were the masses and electrical charges equal, the mathematics of the currently known quarks would be SU(6). But the differences between the quarks are enough that this is probably no longer a good approximation.

However, there *is* a case where SU(3) applies perfectly. It's the theory of the strong force, which is called quantum chromodynamics, or QCD. As we learned in Chapter 2, in QCD, the strong force charges are misleadingly called "color," and there are three colors: red, blue, and green. In QCD theory, one can swap the colors of any two quarks, and it makes no difference at all. Thus, we can say that SU(3) group theory governs the strong nuclear force.

If we can connect group theory to the strong force, can we do that for the other two forces? Yup, although we will forgo a similar motivation between the forces and the groups. The conversation is broadly similar but uses instead the weak and electric charges. The weak force can be described by SU(2) and the electromagnetic force by U(1). ("U" just means "unitary group," which is a technical designation that isn't important here.)

Combining the weak and electromagnetic forces results in SU(2) × U(1), where the "×" means "you need both of these." If we then bring in the strong force, we now have the entire standard model, which can be written in terms of groups as SU(3) × SU(2) × U(1).

For mathematicians and math-minded individuals, writing the standard model this way is ultra-compact and tells a person well versed in group theory everything they need to know. Most of us don't fit into that category though. For us, the much longer, word-oriented, description of Chapter 2 tells exactly the same story.

So why did I go into this conceptually difficult way of talking about the standard model? It's because of how the scientists of the 1970s tried to move the discussion forward. They dove deeply into group theory, hoping to find another and more complex group that would combine the three groups of SU(3) × SU(2) × U(1) into a single, bigger group. How would that work?

Imagine a standard six-sided die used to gamble at places like Las Vegas and Monte Carlo. Suppose the mathematics of SU(3) were based entirely on one side of the die and SU(2) were on the opposite. And, just for fun, U(1) might be inscribed on the sides.

Mathematicians looking at the various sides would see the symmetries and interconnections inscribed on each side, but only by stepping back and seeing the entire die can you see the entire picture and furthermore see that all of the symmetries are part of a bigger whole.

In 1974, American physicists Howard Georgi and Sheldon Glashow realized that the three components of SU(3) × SU(2) × U(1) were contained within a bigger group, called SU(5). If SU(5) applied to the real world, that would be a huge clue toward working out a grand unified theory.

So, you might ask, if SU(5) did apply to the real world, why don't we see that? It's because that SU(5) mathematics would apply at very high energy or, equivalently, very early in the beginning of the universe—shortly after the Big Bang. Then, as the universe expanded and cooled, at a certain temperature, the rules of the universe morphed into the SU(3) × SU(2) × U(1) situation we see today.

Perhaps some analogies are in order. Suppose you're sitting in the dark and you have five railroad cars that seem completely identical to you. Because they're identical, you can swap any two of them, and it makes no difference at all. This situation reflects symmetry in the theory of railroad cars. This is equivalent to SU(5) when the cosmos was young.

Now, at some time, someone turns the lights on, and you see that you have three red and two blue railroad cars. You can now distinguish them, and the situation is a little more complicated. You can still swap the blue cars around and the red cars around, but you can't swap a red and a blue, as that would look different. Continuing the analogy, the instant when the light turns on is equivalent to the universe cooling enough so that what was once a fivefold symmetry is now two symmetries, one threefold and one two. This illuminated situation is equivalent to today's world of SU(3) × SU(2) × U(1).

Here's another analogy. Suppose you have a bottle of 100 proof vodka. Vodka of that proof is 50% water and 50% ethanol. At room temperature, it appears to be a single clear liquid. However, if you cool the bottle below the freezing point of water, you can get slush in the bottle. This is because alcohol freezes at a much lower temperature. In this case, you have an indistinguishable liquid at high temperature but, as you drop the temperature, the two liquids become distinguishable at lower temperature.

This idea of the more general mathematics of SU(5) is great and all, but physics is an empirical science. What matters is whether the theory can make a testable prediction and that the prediction agrees with data. Given that the theoretical community no longer favors SU(5), what tests were done to invalidate the theory?

Well, one consequence is that while the quarks and leptons are very different in the SU(3) × SU(2) × U(1) standard model (remember that SU(3) only applies to quarks), in SU(5), the leptons and quarks are placed on equal footing. The quarks have three strong force colors (red, green, and blue), and the leptons are another color (white). In SU(5), the four colors are equivalent. This means that quarks can swap into leptons and vice versa, and *that* means that it is possible for protons to decay.

How often would that happen? Well, at our current energy and temperature environment, not very often—according to SU(5) theory, it would take, on average, a thousand billion billion billion years for any particular proton to decay. Since the universe is 14 billion years old, that means that protons are, by and large, stable.

However, when you are talking about things like proton decay, statistics is the name of the game. While you may have to wait a very long time if you have just one proton, if you have a thousand billion billion billion protons, you'd have to wait just one year to see one of them decay. And how can you get that many protons? Make a tank containing 3,000 gallons (just over 11,000 liters) of water. And then you wait a year.

Two groups of physicists built such big tanks of water—one in the United States and one in Japan. And, as I said, they waited. And waited. And waited. And they observed no proton decay. And that effectively killed theories based on the mathematics of the SU(5) group.

On the other hand, theoretical physicists and mathematicians know of many more mathematical groups than the few I've mentioned here. For instance, there is another mathematical group called SO(10). SO stands for "special orthogonal," but we don't need to delve into that here. What's important is that in the same way that SU(3) × SU(2) × U(1) is contained in SU(5), SU(5) is contained in SO(10).

Like in SU(5), SO(10) treats quarks and leptons interchangeably, which means that protons would decay. However, in the SO(10) mathematical formulation, the life of protons can be much longer—so long that it would be impossible to build a big enough experiment to test it. So that's kind of where we are. We haven't killed all possible grand unified theories motivated by all possible mathematical groups; however, the enthusiasm for the approach has been tempered somewhat.

Tempered doesn't mean abandoned. For instance, in 2007 American theoretical physicist Garrett Lisi submitted an article to the preprint server arXiv that was written in the spirit of Georgi

Figure 3.4 Figure representing the symmetries and structures inherent in Lisi's E8 theory. While much more complicated than the structure seen in Figure 3.3, it is clearly suggestive of an underlying theory that determines which particles do and don't exist.

and Glashow's paper, but investigating the symmetries and mathematics of a much more complicated mathematical group, called E_8. The "E" comes from "Exceptionally Simple"; however, the group is anything but. Lisi postulated that the group provides a framework for a theory of everything, incorporating the standard model and quantum gravity. The inherent complexity of this group is hinted at in Figure 3.4, which is the E_8 version of Figure 3.3. The proper figure is eight-dimensional, and this is merely a two-dimensional projection.

Lisi's theory has been heavily criticized by the scientific community and has never been published in peer-reviewed literature, so it should be taken with a very large grain of salt. In fact, the paper would never have gotten any attention from the scientific community except that the media heard about it and ran with it first. The media loves Lisi. Although he does have a PhD in theoretical physics, awarded by the University of California, San Diego, he does not have an academic position. He left academia, expressing dissatisfaction with the state of theoretical physics. Instead, he bummed around Hawaii, living in his van, surfing, and doing theoretical physics wherever he happened to be. The media just loved the story—the idea of a rogue, outsider researcher doing work that the established community won't touch is one that Americans love. And, of course, there are many amateur physicists who see Lisi as a kindred spirit—a creative thinker who lives outside the binding strictures of academia.

The fact that Lisi's paper has never been published in a reputable journal, combined with the number of papers that point to flaws in his theory, means that E_8 is not the Holy Grail of physics as it was portrayed in the media. However, it does remind us that symmetries and group theories are still valuable ideas that may well provide hints for those looking to devise a theory of everything.

There is one final thing that the enthusiasm of the 1970s for the group theory approach to search for a grand unified theory can teach us. And this is an explanation for just why theorists think that the energy scale at which the strong force and the electroweak force is unified must be at about 10^{12} trillion electron volts.

In Chapter 2, I talked about the strength of the three well-understood subatomic forces. These three forces have very different behaviors, with electromagnetism having an infinite range, the strong force being limited to the size of a proton, and the weak force having an even smaller range. Because of that, when the

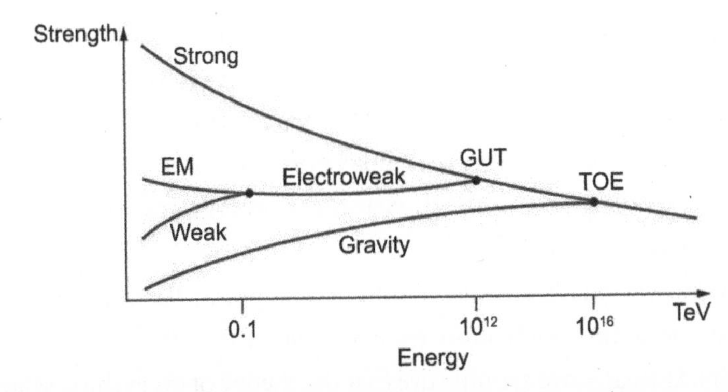

Figure 3.5 The strength of the four known forces varies with energy, with the strong force getting weaker, while the others increase in strength. GUT stands for grand unified theory; TeV stands for trillion electron volts; TOE stands for theory of everything.

strengths of the three forces are compared, a size scale of about the radius of a proton is selected.

However, what I didn't talk about is how the strength of these forces varies, depending on the energy at which they are studied. It turns out that the strength of the strong force decreases as one raises the energy, while the weak and electromagnetic forces become stronger. This was noticed in the mid-1970s by American physicists Howard Georgi, Helen Quinn, and Nobel Prize winner Steven Weinberg.

If the strength of some forces is going up and others are going down, it doesn't take too much thought to realize that they will eventually come to a point where they all will be the same. According to measurements available in the 1970s, the strength of the forces becomes equal at an energy scale of about 10^{12} trillion electron volts. This is shown in Figure 3.5.

More modern measurements are more precise, and they don't quite agree with those of fifty years ago. Current measurements predict that the strength of the forces becomes similar near the energy of 10^{12} trillion electron volts, but not exactly at that value. Even so, that energy scale is interesting. After all, if the strength of the forces is the same, then it is entirely reasonable to imagine that at that energy that the three forces all merge into a grand unified force.

However, that's all speculation at this point. We are not sure that the grand unification energy scale is 10^{12} trillion electron volts. Similarly, we are not sure that the theory of everything scale, where gravity finally joins the party, is 10^{16} trillion electron volts. Both numbers are simply informed speculation, and we will discuss those numbers in more detail a little later.

At this point, the promise that group theory seemed to show in the 1970s seems to have petered out. This isn't to say that it's entirely dead—just that the failure to observe proton decay took some shine off the approach. There are people (like Lisi) who continue to investigate possibilities.

However, even if group theory is no longer as popular of an approach to search for unification as it once was, it doesn't mean that physicists have given up. Instead, they've turned their attention into new directions. In the next two sections, we'll talk about more modern attempts to realize Einstein's dream.

Gravity Gets Loopy

In the last couple of sections, we've talked about some historical attempts to invent a theory of everything or a grand unified theory. It's now time to turn our attention to more modern efforts. As a reminder, a grand unified theory tries to unify the strong nuclear and electroweak forces, while a theory of everything tries to unify a grand unified theory and gravity.

However, that last one—the theory of everything—makes a tacit assumption that needs to be highlighted. Recall that the theories describing the subatomic forces all have a quantum nature, while general relativity is wholly classical, and it has stubbornly resisted all attempts to interject any quantum characteristics. Of course, that's a problem, as before that final theory of everything theory can be devised, physicists will have to work out a quantum theory of gravity. And that's the topic that will occupy us in this section.

General relativity is a hugely successful theory, and it provides a helpful constraint for researchers working on theories of quantum gravity. Specifically, whatever quantum theory is devised, it has to look very much like (perhaps identical to) general relativity when the quantum theory is applied on large distance scales. And, by large distance scales, I mean (ballpark) ten micrometers (10^{-5} m) or larger. That number is set by experimental measurements which have investigated the force of gravity down to about that size scale. The measurements agree perfectly with predictions of Newtonian gravity which, I remind you, agrees with general relativity except in very special circumstances.

So that is very helpful, as it provides theorists a way to check if their explorations of quantum gravity are in the right direction or not.

Another very important hint is provided by general relativity, and that is the very intimate connection between gravity and the geometry of—indeed, the very nature of—spacetime. Understanding that linkage should be a goal of any quantum theory of gravity.

This brings us to a very important distinction. Should quantum gravitational theory be dependent or independent of spacetime? And what does that even mean?

Being dependent on spacetime is pretty easy to understand. Essentially everything you've ever experienced is dependent on spacetime, and that includes all of the known quantum theories. If you ever took an algebra class, you encountered ordered pairs, which are generally written (x,y). The x indicated the horizontal direction, and the y indicated the vertical one. You then would write and manipulate equations with all manner of complicated x and y terms. And, if you were a precocious student, you'd maybe add a third term to the parentheses, for example, (x,y,z), where the z direction would be out of the page of the book.

You might not have realized it, but by writing things in this way, you assumed that space existed—be it two- or three-dimensional. The equations existed in the space, but the space came first.

And space could itself be some odd shape. Restricting ourselves to two dimensions, the shape could be a plane, which is to say flat like a table. It could be spherical, like the surface of a sphere. It could be toroidal, like the surface of a donut. Indeed, it could be something far more complicated. But one begins with the fact that space exists and then the rules of the equations could be worked out.

However, if we're trying to understand the origin of spacetime, then we can't assume that spacetime exists. Instead, spacetime should emerge out of our theory. That's a tall order. How can that happen?

The story begins in the 1800s, when English scientist Michael Faraday introduced the idea of fields to electromagnetic theory. He imagined that electric charges caused electric fields that surrounded them. These fields would then interact with nearby electric charges, and the result would be a force between the two charges.

While fields were useful to calculate forces between objects, the fields themselves had an independent existence. Radio waves are an example of fields that can persist in the absence of nearby charges. For instance, over the last few decades extraterrestrial enthusiasts have attempted to send radio signals to distant stars located many light-years away, which means several tens of trillions of miles. Those radio waves are electromagnetic waves that are traveling in deep space, without the need for additional charges to keep them going. Once they were made, they traveled independently.

While electromagnetic fields travel through space and are therefore dependent on space, it is possible to mathematically construct fields that don't require space to exist. A popular candidate, quantum theory of gravity is based on these space-independent fields. It is called loop quantum gravity, or LQG.

In 1986, Indian theoretical physicist Abhay Ashtekar invented a series of mathematical variables in which Einstein's theory of general relativity could be expressed and which looked more like familiar quantum theory. A couple years after that, Italian Carlo Rovelli and American Lee Smolin used Ashtekar's variables to generate a theory of quantum gravity based on "loops" of fields. These were fields that circulated, kind of like the worm Ouroboros, the mythological serpent from ancient Egypt that bites its own tail.

These loops can intersect with other loops, and the points of intersection are called nodes. The basic idea is illustrated in Figure 3.6. In LQG, the nodes are the smallest bits of space. Their size is thought to be about the Planck length, which, as you will remember from an earlier discussion, is the smallest size for which

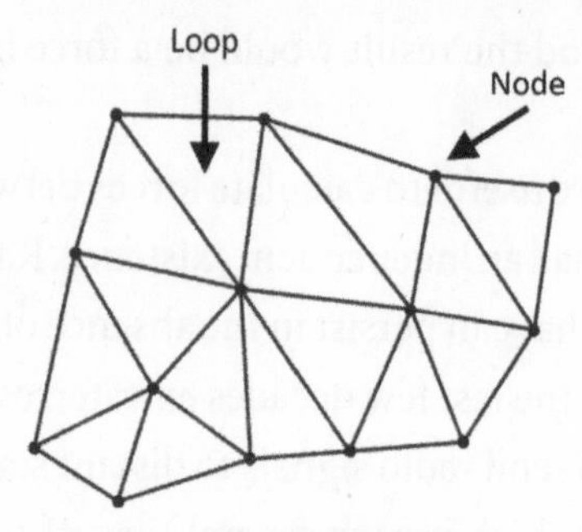

Figure 3.6 Loop quantum gravity is formulated by combining many loops, which intersect with nodes. Taken together, they form what is called a "spin network," and a simple example is illustrated here.

our current theories can apply and below which a new theory must exist.

In LQG, the Planck length is the smallest possible size. Literally space becomes quantized into discrete smallest volumes, like the grains of sand that make up the Sahara Desert. Theoretical physicists love the idea of a smallest size, because it solves an age-old difficulty of many calculations. In those calculations, things like the force or strength of a field increases as the distance gets smaller. At zero distance, the quantity being calculated becomes infinite, which is physically impossible and implies that the theory has broken down. If there is a smallest size that is bigger than zero, then all of those pesky infinities go away and the theorists have far fewer headaches.

If space arises from the nodes between loops, the loops become gravitons. This means that gravitons aren't an intrinsic part of LQG any more than space and time are. They emerge through the interaction of the loops of fields and become apparent only when inspected at larger sizes.

It is important to keep in mind that LQG isn't generally accepted, because there has been no empirical confirmation that it is true. It remains simply a very popular idea. And, given its popularity, it's worth talking about it a bit more.

First, we have to remember that the theory is entirely silent on the behavior of matter and energy within spacetime. The theory is exclusively concerned with spacetime itself. So LQG is most certainly not a theory of everything, nor is it a grand unified theory; however, it may be an important component of both.

Another important feature of LQG is that it doesn't require extra dimensions of space and time—just the three familiar spatial and one temporal one. Nor does it require that there be any symmetries between the matter and force-carrying particles of the known quantum forces. I mention these two points, as they are in stark contrast with the theory of superstrings, which we'll discuss in the next section.

There are, of course, critics of LQG, and one of their potentially most troubling comments revolves around a tension between LQG and Einstein's special theory of relativity.

You may recall that special relativity makes some counterintuitive predictions of how two observers, who are moving with respect to one another, will each perceive a third object as having a different length. Basically, the two observers can't agree on how long things are.

Now take that feature of special relativity and apply it to the size of the quanta of spacetime. If the two observers disagree on the length of objects, then they will disagree on the smallest length of spacetime, and this is a very troubling state of affairs. There are some theoretical physicists who claim that this feature is the death knell of the theory.

As you may know, I make short videos about particle physics and cosmology that I put on the internet, and I made one on LQG.

In it, I mentioned a potential issue with the theory that originates in this claimed tension between special relativity and LQG. I received an email from Carlo Rovelli, one of the architects of LQG, in which he admonished me as not being current on the theoretical situation. He claimed (with good reason and from an expertly informed point of view) that LQG and relativity were entirely compatible, and different observers will not observe quantized spacetime as having different sizes.

On the other hand, there are some of his peers who continue to poke at the theory and are not entirely persuaded by his claim. And the debate will continue.

There is one potential astronomical observation that could test LQG, and it hinges crucially on the existence or nonexistence of the tension between LQG and relativity. You see, if the tension exists, it suggests that different wavelengths of light have slightly different speeds. This is in stark contrast with the classical theory of relativity.

Now, these differences might be very small, which implies that you need to observe objects that are very far away to be able to measure any time differences between the arrival time of light of two different wavelengths. And, when I say, "very far away," I'm talking cosmic distances—ideally hundreds of millions or even billions of light-years away.

Of course, if an object is that far away, it is also extremely difficult to observe. Distant stars appear dimmer than closer ones, just as two identical flashlights will appear to have a different brightness, depending on how far away they are.

Luckily, there is an astronomical phenomenon called a gamma ray burst, or GRB. A GRB is the violent explosion of a star—far more violent than ordinary supernovae. While the details of how

GRBs occur are still poorly known, astronomers have learned enough to determine that most of those that have been detected are extremely far from Earth and furthermore that GRBs are the brightest things in the universe—eclipsed only by the Big Bang itself.

Because of their huge distance from Earth, which could be billions of light-years away, light takes a very long time to travel that distance. If light of different wavelengths travels at speeds that are the tiniest bit different, that means that they should arrive here at different times, and we should be able to see that difference when we detect the light here on Earth. However, measurements have been performed and what a number of gamma ray observatories have found is that light of all wavelengths appears to arrive simultaneously. Thus, the prediction of the speed of light being dependent on the wavelength seems to have been falsified.

If the prediction of the varying speed of light had been an intrinsic prediction of LQG, we could definitively say that the theory has been falsified. However, Rovelli's email reminded me that the reports of the death of LQG are premature.

There are ongoing questions in the study of LQC; for example, it is still not understood how to connect quantum predictions with those of general relativity, but work continues. LQC remains one of the most popular modern efforts to better understand the behavior of gravity in the microrealm and will be so for the foreseeable future.

However, there is competition within the theoretical community for the status of favored child in making progress toward a potential theory of everything. There is one that dispenses with the idea of particles entirely and replaces the smallest frontier of reality with a bunch of wiggling objects called superstrings. Let's turn our attention to this other competitor in the theory of everything Olympics.

A Vibrating Microcosm

For the past half-century or so, superstring theory is considered by many physicists to be the most credible candidate theory of everything. Instead of the quarks and leptons and force-carrying particles of the standard model, superstring theory replaces them with a "string" which, depending on the details of the exact theory, can be imagined by either a subatomic stick of dry spaghetti (which has two ends) or an equally small, tiny hula hoop (where the ends are stuck together). In either case, the known matter and force-carrying particles of the standard model are simply different vibrations—different notes, so to speak—of the string. An up quark might be an A-sharp, while a W-boson might be a C-flat.

It's an ingenious idea, although to be clear, not one that has been proven to be correct. However, given its long popularity, it's one worth investigating in some detail—both its strengths and weaknesses. But first, let me do a brief recap of the history.

One day in 1968, Italian physicist Gabriele Veneziano was studying the strong nuclear force, trying to expand on earlier work by other researchers. This was back in the days before QCD, the modern theory of the strong force, had been invented. He solved the mathematical problem he was working on; however, there was a problem: the solution didn't seem to correspond to reality. There seemed to be no path forward for his work to evolve into a theory of the strong force.

Still, the mathematics was interesting and QCD hadn't been invented yet, so other researchers poked around in the equations. A couple years later and in separate papers, Yoichiro Nambu, Holger Bech Nielsen, and Leonard Susskind came up with a physical interpretation of Veneziano's equations. According to the trio, Veneziano's mathematics accurately described the vibrations of one-dimensional "strings," essentially subatomic sticks of spaghetti. String theory implied that the zoo of particles that had been discovered in the 1950s and 1960s were just different vibrations of the strings. This conclusion turned out to not be correct, but some of the ideas were to return a decade and a half later.

Veneziano's work only covered particles which had spin that was an integer multiple of $\hbar$, which we remember is called the reduced Planck constant. Nowadays, particles with that category of spins are the force-carrying particles of the standard model—the photon, gluon, W and Z bosons, and the Higgs boson. However, in 1971, Pierre Ramond generalized Veneziano's work to include particles with spin that is a half-integer multiple of $\hbar$. The fundamental particles that we know of today with that property are the quarks and leptons. Again, this was all before the standard model was really developed and properly understood, so one must be very cautious when understanding what had been accomplished. That's because we know more than they did back then. Ramond had generalized the mathematics so that it could describe what we now call matter and force-carrying particles, although at the time it was still being investigated as a possible theory of the strong force. However, shortly after this time, QCD was invented and Veneziano and Ramond's mathematics were discarded as a candidate theory of the strong force.

That could have been the death knell of string theory, but in 1974 John Schwarz and Joël Scherk noticed that Veneziano's mathematics predicted a vibrating string with exactly the properties of the graviton, which is a hypothetical particle that transmits the force of gravity. Gravitons have not been observed and may not exist, but we can use known physics to work out their properties if they do. Because gravity has infinite range, gravitons must have zero mass—just like photons and for the same reason. Because gravity neither absorbs nor emits light, gravitons must have no electrical charge. And because of the equations of general relativity, if gravitons exist, they must have a subatomic spin of 2 units of $\hbar$. Schwarz and Scherk found that string theory predicted a vibration pattern which had these properties.

String theory started out as a candidate for the strong force, but with the discovery that it could predict a possible particle that was a graviton, Schwarz and Scherk began to claim that earlier researchers had been wrong all along. What Veneziano had started wasn't a theory of a strong force, but rather Einstein's Holy Grail. They began to suspect that string theory was actually a theory of everything.

I've talked about string theory; however, the candidate theory of everything is called superstring theory. What is it that makes superstring theory . . . well . . . super?

In the early 1970s, a small army of theoretical physicists were exploring the mathematics of quantum electrodynamics—the quantum version of electromagnetism—trying to find ways to build a similar version of the strong force. This was before the days of QCD, which is currently our theory of the strong nuclear force. And, given that nobody knew the answer yet, researchers were exploring many ideas.

I've talked about symmetries within the context of physics theories, although I have barely scratched the surface of the subject. Of the many known symmetries, one that we've encountered is electric charge, where we can swap what we call positive and what we call negative, and the equations of electromagnetism are essentially unchanged. There is also the symmetry of the strong force, where the red, blue, and green charges can be swapped without changing the theory.

In the early 1970s, a new hypothetical symmetry was proposed. It proposed that maybe the equations should be unchanged if we swapped the force-carrying particles and the matter particles. This possible symmetry is called "supersymmetry."

That just sounds murky, but it's actually a very simple thing. However, to understand it, we need just a few new concepts. All force-carrying particles have a subatomic spin that is an integer multiple of $\hbar$. Particles with an integer spin are called "bosons." All matter particles have a subatomic spin that is a half-integer multiple of $\hbar$. Particles with this property are called "fermions."

So let's imagine some sort of equation that includes both fermions and bosons. Let's call fermions "F" and bosons "B." A simple possible equation might be as follows: (original) = $F^3 + B^2$. (Note that this equation has nothing to do with physics. I just made it up to illustrate my point.)

If we swapped fermions and bosons, which we do by swapping the F and the B, we get a new equation: (swapped) = $B^3 + F^2$. We see that the original and swapped equations are distinguishable. Consequently, we would say that our equation is *not* supersymmetric, because they are different.

Now suppose that we have a different initial equation: (original) = F + B. If we swap F and B, we get this: (swapped) = B + F. These two equations are the same, just like 3 + 4 = 4 + 3. Such an equation is therefore supersymmetric.

So that's all there is to the idea of supersymmetry. It just means that if you swap all of the F's and B's in the equation that it looks exactly the same. And if you make an equation describing strings that has this swapping property, that means that you now have a superstring theory.

In 1984, superstring theory really took off. By that time, superstring theorists were beginning to think of the familiar quarks, leptons, and force-carrying particles of the standard model as just different vibrational patterns of a single string. Furthermore, researchers could begin to understand how the mathematics of superstring theory could—at least in principle—map both onto the standard model and quantum gravity. This sounds hugely promising as a potential theory of everything.

The mathematics of superstring theory is very difficult and when one explores it, the theory has some surprising consequences. One thing that is hard for some to accept is that the theory fails badly when it is written in the standard three-dimensional space that we are familiar with. However, the theory works just fine if there are more dimensions added to it. In order for superstring theory to work, we need not three space dimensions, but rather nine. And, of course, if you add in time, then superstring theory needs a ten-dimensional spacetime to work.

Of course, it certainly looks like requiring a ten-dimensional spacetime is a fatal flaw of the theory and we should discard it. I mean . . . after all . . . space is obviously three-dimensional. However, we shouldn't be hasty. You may recall Einstein attempt to unify gravity and electromagnetism and that Theodor Kaluza was able to do that if he wrote his equations in a five-dimensional spacetime. Oskar Klein then showed how the extra dimension of spacetime could be very small, which was absolutely necessary for Kaluza's theory to have a chance.

Similarly, in superstring theory, physicists postulate that of the ten dimensions of spacetime, four are the ones with which we are familiar and the other six are tiny. Tiny dimensions are very hard to imagine, but we can remember the tightrope walker analogy. Viewed from a distance, it appears that the tightrope walker can walk in only one dimension—forward and backward is the only option. But if you shrink yourself down to the size of a small ant sitting on the rope, the situation is different. The ant sees the long dimension we do, but it also can walk around the rope. That's a small second dimension.

For superstring theory, you just have to imagine that at each spot in our familiar four-dimensional spacetime that there are six additional tiny dimensions that we can't see. And, if that's true, then superstring theory is possible. If it's not, then superstring theory is merely a cool idea that ultimately didn't work out.

The size of those extra dimensions is thought to be somewhere in the neighborhood of the Planck length. If that's so, we have no equipment capable of looking to see if they're real or not. Because of that, the jury is out on the existence of those extra dimensions.

So that's a quick sketch of what superstring theory says. Instead of the twelve matter particles (six quarks, three charged leptons, and three neutrinos) and seven force-carrying particles (photon, gluon, two W's and a Z, the Higgs boson, and the hypothetical graviton), there is only a single string, with different vibration patterns corresponding to each of the particles of the standard model. And the vibrations exist in a ten-dimensional spacetime. Furthermore—and this is crucial—the theory can bring together the known subatomic forces together with a quantum theory of gravity. Not bad.

There's only one problem. We can't test the theory to see if it is true. Superstring theory, if it's real, works at energies near the Planck energy, which is a quadrillion times higher than we can create in our laboratories. So we experimenters cannot build equipment to test the theory in its natural regime. And theorists are unable to make predictions using their equations at an energy scale that we can test. So we're stuck. There is no known way to test superstring theory. And an untested theory should not be believed.

Even worse, within superstring theory there are many parameters that need to be determined. Since we don't know the value of those parameters, superstring theory predicts about 10^{500} different possible configurations. As if the disparity between the energies we can measure and the ones in which superstring theory weren't enough, having such a large number of possible configurations just makes things incomparably worse.

There is much more that one can say about superstring theory, and the Suggested Reading contains many delightful hours of learning if you really want to dig into it. However, for our purposes, it is sufficient to know merely the basics and then understand that the theory will not be testable in the foreseeable future.

Superstring theory has been with us in one form or another for nearly half a century, without a single, solid, testable, prediction, beyond the idea that extra dimensions exist. This has led a number of physicists to become disenchanted with the entire idea. Books have been written about the limitations of superstring theory, and those are in the Suggested Reading as well.

Superstrings and LQC are the two most researched theories as physicists look for what comes next after the standard model. It's probably worth comparing and contrasting the two.

As mentioned in the previous section, LQC is not a candidate theory of everything. It is merely a theory of quantum gravity. It begins with the idea of fields, and spacetime is a consequence of the theory. It predicts that spacetime is quantized, meaning that it comes in discrete chunks. Furthermore, in its common form, it only requires the familiar four dimensions of spacetime. And, unlike superstrings, supersymmetry is not part of the theory.

In contrast, superstring theory is not a theory of spacetime. A smooth and continuous (e.g., not quantized) spacetime is assumed. Thus, the theory is silent on the origins of space and time. It is a candidate theory of everything, and it requires that extra dimensions of spacetime exist and supersymmetry is an intrinsic part of the theory.

Both theories suffer from a lack of empirical validation, mostly because they describe an energy realm beyond what we can currently create in the laboratory. While both theories are respectable among the community of professional physicists, they remain, for the moment, incomplete. So where does that leave us?

What Now?

For the past several sections, we have been learning about a number of historical and current attempts to uncover a theory of everything. These attempts have spanned nearly a century. None have been successful.

English is full of aphorisms that often contradict one another. One that might be appropriate here is "If first you don't succeed, try, try again." And that seems like good advice. History is replete with tales of people who have encountered obstacles but, with grit and determination, have persisted and achieved their goals.

On the other hand, a contradictory aphorism is one often misattributed to Einstein, "Doing the same thing over and over again and expecting different results is a sign of insanity." In our situation, it is perhaps valuable to ask if the failed approaches of the past are a sensible path to follow as we continue the search.

Given that there is a disagreement in the kind of folksy advice one might get, perhaps the best choice is to step back and take stock of where we've been and where we still want to go. Let's do that.

Past attempts to develop either a theory of everything, or even just a grand unified theory, have been entirely within the province of theoretical physics. There is merit in this approach, as scientists can let their imaginations run wild and speculate freely. Any idea not yet ruled out by existing data is still a valid path of inquiry. This approach is closely aligned with the mythology that has arisen around Einstein—a great thinker, the story goes, who, by sheer brilliance, worked out his theory of special relativity as he calmly puffed his pipe in his study.

And there is a grain of truth in that mythology. Einstein was fond of what he called "gedanken experiments," thought experiments whereby he simply imagined situations and worked out his theories. And there is no doubt that he had success with that approach. However, we should not forget the last three decades of his life, where this technique was thoroughly unsuccessful. He developed tunnel vision, concentrating on electromagnetism and gravity, with an occasional (often insightful) critique of quantum mechanics. He ignored the burgeoning world of newly discovered particles and the discovery of both the strong and weak nuclear forces.

Einstein's introspective techniques aren't unique. For instance, there is also the success of the theoretical community in the 1960s and 1970s, where the idea of symmetries produced both the theory of quarks and the equations that resulted in electroweak theory.

However, it has been about fifty years since progress has been made beyond the standard model, which could be interpreted as telling us that the techniques of the 1960s are not the way of the future. And then there's the principle of supersymmetry, the requirement that fermions and bosons are interchangeable in a theory's equations. It was all the rage since the late 1980s and, even as late as the early 2010s, the LHC was thought to be the accelerator that would validate that theory. The LHC began operations in 2011 and, over a decade later, searches for supersymmetry using that facility have come up dry. That is not to say that the LHC research program has been useless—after all, scientists using the facility discovered the Higgs boson and have published over three thousand papers as of 2022. However, as far as supersymmetry is concerned, the research has come up empty handed.

It may be that the theoretical approach of using symmetries and a perceived beauty of equations as a guide star for the future has run its course. Iconoclast Sabine Hossenfelder has written a book called *Lost in Math,* in which she criticizes the theoretical community and, by extension, the entire high-energy physics research community at large for its attachment to techniques that were successful in the past but are no longer. Personally, I think she paints with an overly broad brush and unfairly generalizes the sins of a few to the broader community. However, there is no question that there are those who cling to symmetry as a productive well to which they return again and again, even though there have been no recent successes.

Where does that leave us? It is perhaps worth taking a huge step back and remembering that physics is ultimately an experimental endeavor. The most lustrous and compelling theory can be killed by inconvenient facts.

So let's recap a bit with what we know.

We know that the strength of the subatomic forces differs as the energy at which they are tested changes, with some increasing in strength, while others decrease. And furthermore, if we project the trends we've observed, we see that in the neighborhood of an energy of 10^{12} trillion electron volts, that the strength of the electroweak and strong force becomes comparable.

We know that the combination of our current understanding of quantum mechanics and general relativity tells us that at the Planck energy, about 10^{16} trillion electron volts, general relativity fails totally and must be replaced by an improved theory.

We know that, using the LHC—the world's most powerful particle accelerator—we can collide protons together at near the speed of light, achieving energies of about 10 trillion electron volts.

We know that general relativity fails when applied to the quantum realm, predicting infinities, and therefore there must be a new and improved theory, called quantum gravity, that describes the behavior of the world in the microcosm. This insight places no constraints on the energy at which quantum gravity must become important.

We know that the universe is currently expanding uniformly, and this observation is consistent with what we call the theory of the Big Bang. Running the clock backward, we know that the universe was once smaller and hotter and that in the early times the universe was very different. We also know that the field of particle physics studies the behavior of matter at high temperatures and energies and that this gives us a good idea as to what the conditions of the early universe were.

We also know that scientists have built two very successful and compelling theories that describe much of what we see in the world around us—the standard model and the theory of general relativity. These ideas have shown the connected origins of phenomena that seem to have nothing to do with one another.

We know that we've tested the standard model from sizes as small as about 1/10,000 the size of a proton to perhaps ten times the size of the proton (the strong force), to atomic sizes (the weak force), and the size of the universe (electromagnetism). We have tested all of these forces at temperatures ranging from the coldest possible temperature (−273.15 °C, −459.67 °F) to as hot as 5.5 trillion degrees Celsius (9.9 trillion degrees Fahrenheit).

We know that general relativity successfully predicts the behavior of gravity from a size scale of about 1/100 a millimeter to much larger than the solar system. We know that, with a few caveats, the theory applies to the visible universe. We'll return to those caveats in subsequent chapters.

Then there are the things that we don't know, but we merely think.

We think that at about 10^{12} trillion electron volts that the known subatomic forces merge and will be shown to be different facets of a more complete theory that we call the grand unified theory.

We think that at an energy of about 10^{16} trillion electron volts that the grand unified theory and the theory of quantum gravity will be shown to be two special cases of a single theory: the theory of everything.

Figure 3.7 illustrates what we know and what we think. The solid lines represent well-understood connections between various

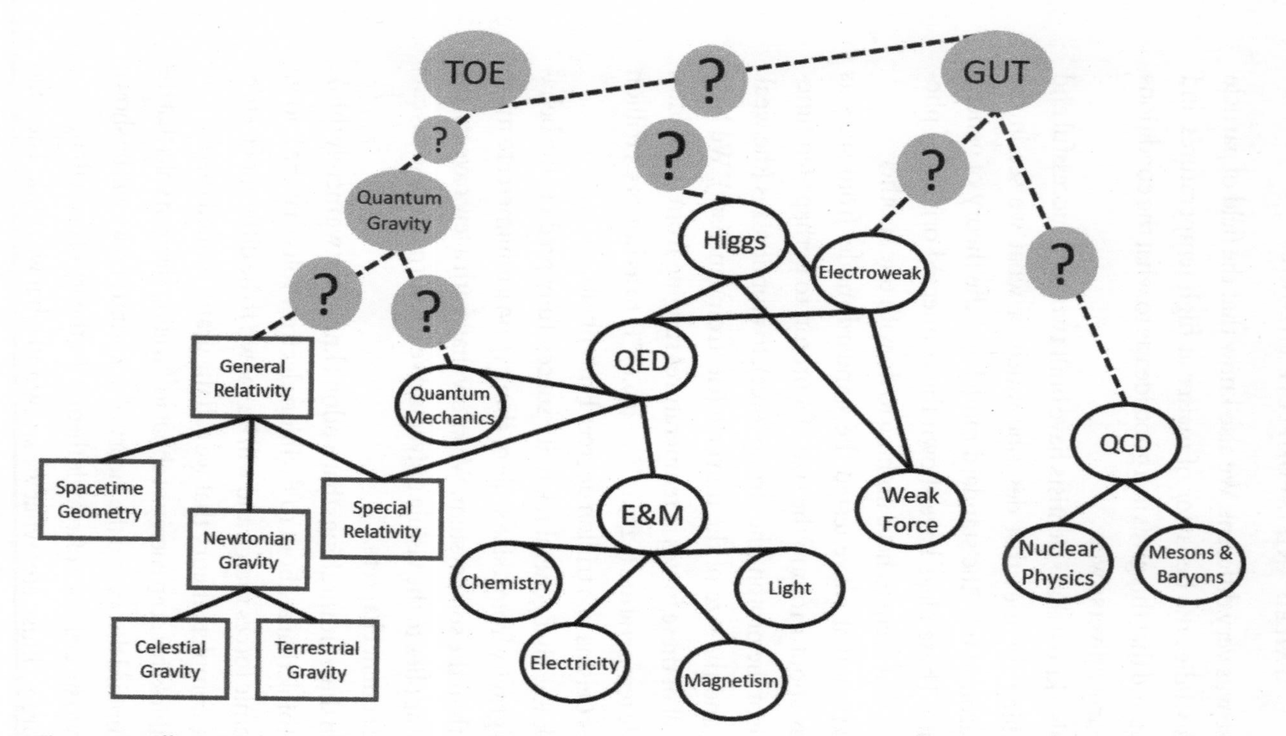

Figure 3.7 Illustration of the known and suspected connections between various topics. White ellipses denote the standard model, rectangles denote gravitation, the gray ellipses are speculative unifications, and circles with question marks illustrate connections for which we are ignorant. Finally, solid lines represent well-known connections, while the dashed ones represent speculative ones.

phenomena, while the dashed ones represent connections that have not yet been proven. It's a delightful graph which suggests that we're well on our way toward working out a theory of everything.

But we shouldn't be hasty. We need to take a hard look at a few numbers. It is often said that the energy scale of a grand unified theory is 10^{12} trillion electron volts and the energy scale of a theory of everything is 10^{16} trillion electron volts. In contrast, the highest energy that we can create on Earth and study is a mere 10 trillion electron volts. This means that the energy at which we expect a theory of everything to apply is 10^{15} times higher than we can study today. That's a quadrillion times higher. And, to quote a precocious toddler, a quadrillion is a lot. It's not only possible but also very likely that there we will encounter many phenomena that we can't even imagine as we attempt to achieve energies closer and closer to the Planck energy.

Most people don't have a quantitative sense of energy, so let's use an analogy involving length to get a sense of how a change of a scale of a quadrillion can result in some astonishing surprises.

Suppose you were one of our early ancestors—say the *Australopithecus afarensis*, who is thought to have left the oldest known hominin footprints at Laetoli, Tanzania, in Africa. Let's say that the reach of your arms is about a meter. That's an overestimate, but it's in the right ballpark for our purposes. What are the sorts of things you are familiar with?

Laetoli is located in northern Tanzania. It is dry and rough terrain, although it was wetter in the days when *Australopithecus afarensis* strode the land. Olduvai Gorge, where the fossil Lucy was found, is some 50 km to the north, Ngorongoro Crater is about the same distance to the east, and Lake Eyasi is about half that distance to the southeast. Lake Victoria is a distant 300 km away to the northwest, and Mount Kilimanjaro is a similar distance to the northeast. Stretching to the west for much longer distances is a flat plane that is seemingly endless.

Let's assume that you must wander to collect food and that your range is 100 km in every direction. That's 100,000 (10^5) times your arm span. As you travel, you take stock of the world around you. The terrain is generally not all that hilly, with altitude changes of about a couple hundred meters. There are one or two mountains with peaks that rise to a height of about 1,500 meters. You will encounter dry riverbeds and Lake Eyasi, which has an area of about 1,000 square kilometers. Your travels aren't enough to see Mount Kilimanjaro, nor Lake Victoria, so your experiences won't include them.

Laetoli is near the equator, so it's always hot. The terrain is dry part of the year, and there are torrential rainstorms in the spring.

So that's your world—generally hot, with wet and dry periods; modest altitude variations, with an occasional peak rising above the plane; there are riverbeds that are occasionally full of water and one large lake; and, of course, there are the flora and fauna that you encounter in your journeys.

If you have these experiences and you were to try to project out what the world must be at distances a quadrillion meters from you, you'd likely expect more of the same; after all, what you've encountered is most likely what you'll expect. You might imagine slightly bigger mountains or a larger lake, but to envision something completely different is just uncommon.

So let's extend the distance tenfold, to a thousand kilometers, which 10^6 or a mere million meters. On distance scales like this, you will encounter the Indian Ocean, something you probably never would have imagined. The water seems to go on forever.

It's salty. And it's inhabited by giant "fish," whales and sharks that could swallow you whole. With this increase in distance, you've encountered your first true surprise.

And then, of course, to the northwest, you encounter the rainforests of the Congo. A world where it's wet nearly all of the time maybe isn't quite as much of a surprise as the ocean, but it's still something unexpected.

Let's up the distance another factor of ten—this time to 10 million (10^7) meters. Now you've discovered Antarctica, a frigid wasteland of frozen ice, inhabited only by penguins—flightless birds that hunt underwater for long periods of time. This is something you'd never have imagined in your wildest dreams. The very concept of ice would be foreign to you, unless you had previously happened on the glaciers of Mount Kilimanjaro. And you'll encounter the European Alps, a vast swath of mountainous terrain unlike anything you've ever encountered. There's also the Sahara Desert—an endless tract of nothing but constantly shifting sand and no plants, animals, water, nothing.

I hope you're getting the idea. Another factor of ten will encompass the entire world, with all of the variety of the Earth. You'd encounter the tornadoes of the American Midwest, something that is a rarity in your native land. Kangaroos in Australia, giant squid in the deepest oceans, indeed, locations at the bottom of the oceans where light never penetrates—these are all things that you'd never have predicted if you tried to project from your experience of the world within arm's reach to a size that is a quadrillion times larger.

And just so we understand the true extent to which you'd be surprised, I haven't mentioned the existence of outer space just a mere 100,000 (10^5) meters straight upward. You'd be completely surprised by the existence of entirely different worlds—the hellish atmosphere of Venus, the elegant rings of Saturn, and the airless and sunbaked surfaces of the Moon and Mercury.

Indeed, when the modern you asks what objects are within a sphere with a radius of 10^{15} meters, you find that it is a tenth of a light-year. That sphere encompasses the entire solar system out to the inner edges of the Oort cloud, the buzzing swarm of frozen bodies that occasionally plunge into the solar system as a spectacular comet that you might have seen light up your night sky.

In short, projecting from the realm that you know to something a quadrillion times bigger is a bit foolish. It is almost a certainty that you will encounter many unexpected phenomena, and your theory will need to be revised and updated and even discarded and remade again and again.

I hope this analogy has really driven home the enormity of the challenge we are facing. Surprises are inevitable, just as researchers of the late nineteenth and early twentieth century were amazed when X-rays and radioactivity were discovered. They were astounded by the fact that the supposedly uncuttable and fundamental *atomos* of ancient Greek is really the complex structure that we understand modern atoms to be. And, of course, we can't forget the strong and weak nuclear forces of the standard model, which were unimagined in the late 1800s, and all of Einstein's mind-blowing insights as to the nature of space and time. We've been surprised again and again as we try to understand the universe around us.

So where does that leave us? Is it simply a fool's errand to try to search for a theory of everything?

Well, no. It's not. But the conceit that a theory of everything will just pop into the mind of some starry-eyed genius as they stare off into space is just that—a conceit. It simply beggars the mind that anyone could anticipate everything that exists between today's research frontier and the distant Planck energy scale.

If I've dashed your hopes of a quick answer to this grandest of questions, I do not want to give you the impression that the aim of fundamental physics and cosmology is hopeless. After all, we've made a great deal of progress since the time of the ancient Greeks, with the vast majority of discoveries occurring in the last half a millennium. The origin of modern science can be identified as 1543, the watershed year where Polish polymath Nicolaus Copernicus published his *On the Revolutions of the Heavenly Spheres,* where he proposed that the Sun and not the Earth was the center of the universe.

In the five hundred years since Copernicus's opus was published, we've gone from a world in which learned scholars debated the location of the Earth in the cosmos, to knowing about a vast and expanding universe, full of billions and billions of stars. We've learned a lot and we understand a great deal of all manners of science.

How did that happen? Discoveries. Scientists have looked at the world around us, poking in this corner and that, finding clues that have taught us of the world of atoms and galaxies. With each discovery, we've had to rewrite our books with an ever-improving level of understanding. And that is how we'll continue to move forward. Discoveries will guide theoretical physicists and cosmologists who will slowly piece together the puzzle, taking step after painstaking step toward their final goal.

So that brings us to the real question: What is the next step? And that is where we are in great shape. Observations that we don't understand are clues—hints on how we can improve our understanding. And there are many such clues.

Figure 3.8 revisits an earlier figure, with its web of things we now understand and hopes for future progress. However, surrounding

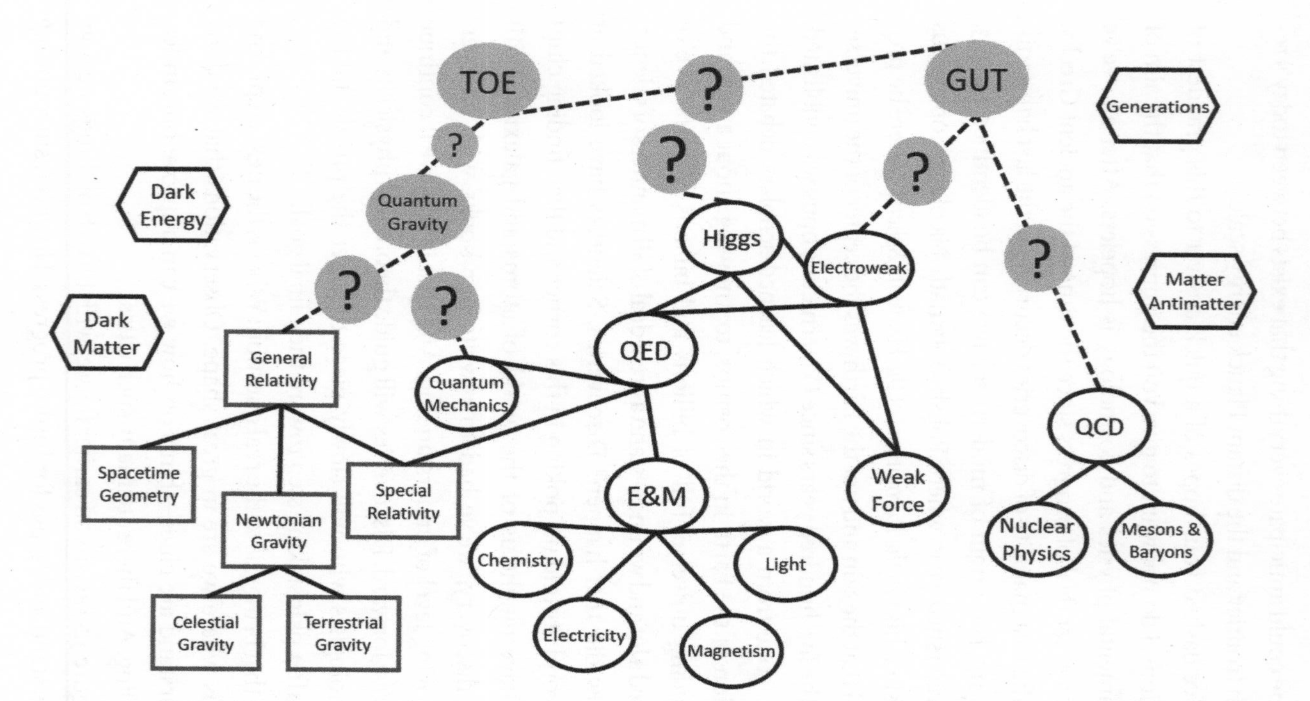

Figure 3.8 Known and suspected connections between known physical phenomena. The hexagons represent phenomena that have not yet been successfully placed in a successful theory of everything.

that intellectual edifice are mysteries—scientific observations that puzzle today's researchers. And here's the most exciting thing—we don't know how any of these fit into our original web. We will spend the rest of the book understanding these mysteries in more detail, but here we can breezily mention one just to get a flavor of how the future will unfold.

Dark matter is one of the mysteries that we don't yet comprehend. Observations of how galaxies rotate don't agree with predictions, and we don't truly understand why. One possible explanation is that there exists a form of undiscovered matter that permeates the universe and solves the mystery. If that's so, then the dark matter hexagon of Figure 3.8 will fit somewhere in the web over in the section that houses the strong and electroweak forces. On the other hand, if the answer lies in the fact that we don't understand gravity well, then the dark matter hexagon rightfully belongs over on the left-hand side, somewhere in the realm of gravity.

And this is how it will be for all of the hexagons that are unattached. As we learn more about them and understand their nature, they will fit somewhere in the web, perhaps creating new branches and pointing to connections that we haven't even imagined. The simple web of Figure 3.8 proposes connections from the world that we know to the speculative worlds of grand unified theories, quantum gravity, and a theory of everything. These connections might need to be heavily modified, and the final result may look very different than we now imagine.

That's the truly exciting thing about doing fundamental physics research—we have only the haziest idea of the proper path forward. We will continue to be puzzled and stymied, only to be delighted when we have an insight that clarifies another bit of the web.

Our goal hasn't changed, but now we have a much better appreciation of the daunting magnitude of the task that lies ahead of us.

Still, as they say, the journey of a thousand miles begins with a single step. We're well beyond the first step, but countless steps lie before us, beckoning us with a world of wonder. In the following chapters, we'll talk about the mysteries that we're trying to solve now, hoping to further humanity's long journey toward that elusive theory of everything.

CHAPTER 4

DARK MATTER

In the last chapter, Figure 3.8 laid out the known connections between observed phenomena, as well as several mysteries that need to be solved. Figuring out the explanation for those mysteries and how they fit into the web of what we understand will be a valuable step forward in our attempt to get closer to a theory of everything.

But, of course, we don't know the answer for any of the mysteries yet, and they might fit in anywhere in the existing web, or they may even mean that there are additional web connections that we don't yet even know about. So, in the next few chapters, I'm going to dig into what we know and what we hypothesize about where these mysteries fit in. By the end, we'll have a good understanding of the possible ways our picture of the cosmos and the laws of nature might improve in the next few years or decades.

Our first mystery is dark matter. Dark matter is thought to be a form of matter that doesn't absorb or emit light or interact electromagnetically in any way. It is also not affected by the strong or weak nuclear forces. Perhaps most dramatically, it is about five times more prevalent than the ordinary matter, and it permeates the entire cosmos. And, of course, if there is five times more dark matter than ordinary matter, it means it is imperative that we know

Einstein's Unfinished Dream. Don Lincoln, Oxford University Press. © Oxford University Press 2023.
DOI: 10.1093/oso/9780197638033.003.0004

how it fits into our web—indeed, purely from the point of view of total mass—understanding dark matter is more important than ordinary matter; after all, there's more of it.

I just told you about the properties of dark matter, and I even gave some numbers, which means that we are at least certain it exists, right? Actually, that's not really true. At the present time, researchers are not 100% certain that dark matter is real. Dark matter is a name we give to one possible solution to a bunch of astronomical mysteries. In fact, dark matter might not exist at all.

With this level of uncertainty, we should proceed with a certain amount of caution and review what we know, what we don't, and what could be possible. Let's start the way all good stories do—at the beginning.

Fritz Zwicky was a Swiss cosmologist who spent the bulk of his career working at the California Institute of Technology. In 1925, he received a prestigious fellowship from the Rockefeller Foundation which allowed him to migrate to the United States to work with legendary physicist Robert Millikan.

Zwicky was a genius, but he was also a difficult individual—irascible and combative. He frequently argued with his colleagues and even his friends found him to be prickly. He is famous for calling those who crossed him "spherical bastards," because they were bastards no matter what way you looked at them.

In 1933, Zwicky was studying the Coma cluster of galaxies. Galaxy clusters are just what the name suggests, large groups of galaxies that are near enough to one another to be held together by gravity. This means that the galaxies stick relatively close to one another, like the planets around our Sun, rather than simply flying by one another in sort of a chance encounter.

Of course, "close enough" in astronomical terms is quite large. The Coma cluster is about 20 million light-years across, and it is comprised of over a thousand individual galaxies. It is located a little over 300 million light-years away in the constellation Coma Berenices, which is just a few degrees away from the galactic north pole.

Zwicky was studying how gravity held the cluster together. By looking at the brightness of the galaxies, he could figure out how many stars were in them. And, by knowing how much stars weigh, he could then figure out how much mass was in the entire cluster.

He could also use what is called the Doppler effect to figure out the speed of the galaxies. The Doppler shift just means that objects moving away from an observer are redder than they would appear if they were stationary. And if the objects move toward an observer, they appear bluer.

Since stars predominantly are big balls of glowing hydrogen gas, and he knew what colors were emitted by hydrogen, he could look at the colors of the galaxies of the Coma cluster, see by how much they looked bluer or redder than they would if they were stationary, and work out their speeds.

By knowing the galaxies' masses and locations, he could work out the gravitational forces holding the galaxies together, and he could compare that to their speeds. In order for the cluster to stay together, the galaxies had to be moving slowly enough that they were caught in the cluster's gravitational pull.

What Zwicky found surprised him. He found that the galaxies were moving far too fast to stick together. By all rights, the Coma cluster shouldn't be a cluster at all. Instead, the galaxies should be shooting willy-nilly all over the cosmos.

To explain his observation, Zwicky had to come up with a couple of hypotheses. One was that the Coma cluster really wasn't a cluster; instead, it was a bunch of high-speed galaxies that happened to be in the same place. That seemed unlikely, and today's scientists have ruled it out.

Another hypothesis was that the galaxy cluster simply had more mass than he could see using his telescopes. And we're talking a lot of missing mass. According to Zwicky's 1933 paper, the Coma cluster needed to be 400 times heavier than his measurements suggested just to exist, but his initial observation was based on the combination of only seven measurements, drawn from averages of up to sixty galaxies. And modern measurements, incorporating the observation of over a thousand galaxies, come to a similar conclusion as Zwicky's, with the more modest estimate of the mass of the Coma cluster as ten times greater than the matter that can be seen with today's optical and radio telescopes.

Zwicky called his hypothesized and unseen matter "dunkle Materie," or dark matter. At the time, dark matter was simply defined as matter that didn't give off light, although modern definitions are considerably more precise. According to Zwicky, there was literally more to the universe than meets the eye.

Zwicky's observation and hypothesis were an unsolved and somewhat forgotten mystery for about forty years. And then an inquisitive astronomer made a perplexing measurement.

Spinning Galaxies

Vera Rubin was born in 1928, and she showed an early interest in astronomy. As a young child, she built crude telescopes with her father. She studied astronomy in college, receiving a bachelor's degree from Vassar College and eventually her PhD in 1954 from Georgetown University in Washington, DC.

As part of her PhD work, she concluded that galaxies often clumped together in large groups. In the decade following her degree, she did additional work where she claimed that the universe wasn't stationary; rather, galaxies were carried along together in large flows, as if the galaxies were floating in a vast cosmic river. Although both of these conclusions have been subsequently confirmed, they were highly controversial at the time, and she encountered a significant amount of resistance.

Wishing for a respite from the hostile environment she encountered, she turned her attention to a safe and seemingly noncontroversial topic: the rotation speed of galaxies. In collaboration with Kent Ford, an instrument maker of some note at the time, she began by looking at the Andromeda galaxy. Because of the galaxy's rotation and the Doppler shift, one side of the galaxy appeared bluer than if it wasn't rotating, and the other side looked redder.

Rubin was pretty sure she knew what she'd see. As far back as 1619, German astronomer Johannes Kepler predicted the orbital speed of moons around planets and planets around stars. Stars around galaxies should be identical. And, of course, Isaac Newton greatly clarified what was going on with his laws of motion and gravity. The derivation is super easy if you've had a single semester of algebra-based physics, and a reference of how it is done is given in the Suggested Reading.

Figure 4.1 shows what Rubin expected. She expected that stars near the center of the galaxy would orbit slowly and that stars orbiting at larger radii would move more quickly. Then, in the very outskirts of the galaxy, stars would move slower and slower.

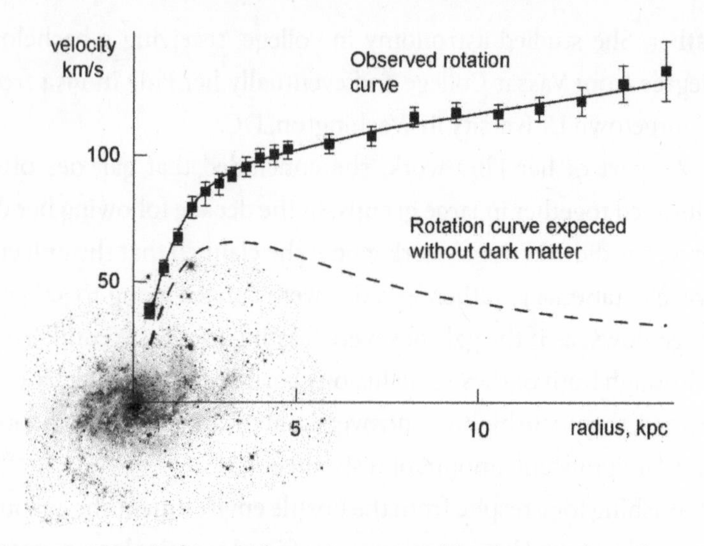

Figure 4.1 Vera Rubin used rotation curves like these to show that galaxies rotate faster than the laws of physics and observed amount of matter can explain. The galaxy is the Triangulum galaxy.

This pattern—speeding up and then slowing down as one gets farther and farther away from the center of the galaxy—originates from the fact that only mass inside of an orbit will affect the orbital speed. Mass outside the orbit has no effect at all. And for orbits well inside the galaxy, as you make an orbit bigger, that orbit encompasses more mass. When you get to the very outskirts (or even outside) of the galaxy, further increases in the orbit do not encompass more mass. That's when the orbital speed of stars starts to drop again.

So what did she see? As you can see in Figure 4.1, near the center of the galaxy, the measurement and prediction agree pretty well. However, when one reaches the outskirts of the galaxy, the prediction is very clearly wrong. She published her first paper describing measurements of the Andromeda galaxy in 1970 and a follow-on paper in 1980 that repeated the measurement for twenty-one galaxies at a variety of distances from the Earth.

We now have two mysteries that tell us that there is something awry with our understanding of the cosmos. Before we talk about possible solutions, let's explore one more unexplained astronomical mystery.

A Cosmic Magnifying Glass

In Chapter 2, we talked about Sir Arthur Eddington's 1919 observation of the location of stars in the Hyades cluster and how they were distorted by the passage of light near the sun. You'll recall that the phrase that describes the phenomenon is called gravitational lensing.

However, gravitational lensing can leave a more dramatic impact on the sky than slightly changing the apparent location of a handful of stars.

In 1924, Russian astronomer Orest Khvolson was playing around with Einstein's general relativity equations, when he realized that if an observer and two stars lay on a straight line, that light from the more distant star would be bent by the closer one, and the result would be that the observer would see a "ring" around the closer star—the distorted image of the distant one. Such rings are called Einstein rings or, occasionally, Einstein-Khvolson rings.

It turns out that the size of a such a circle created by the gravitational field caused by a single star is very small—in fact, it is too small to have been observed even by the most precise modern telescopes. However, not only stars can deflect light. Galaxies can, too. In this case, the distances are greater, and the gravitational fields are stronger and seeing such a phenomenon is possible.

In 1998, a team of astronomers combined data from the British Merlin radio telescope facility and the Hubble Space telescope, and reported the first image of an Einstein ring. JVAS B1938 + 666 is a giant elliptical galaxy, located about 7 billion light-years from Earth. When the researchers imaged the galaxy, they saw that it was surrounded by a ring—the distorted image of an even more distant galaxy. Since then, many Einstein rings have been observed, and an example is shown in Figure 4.2.

While the observation of Einstein rings was certainly of considerable interest, it turns out that galaxies aren't the most massive objects in the universe. Even larger objects are clusters or

Figure 4.2 The galaxy LRG-3-757 is an example of a nearly perfect Einstein ring. (Figure courtesy ESA/Hubble/NASA.)

superclusters of galaxies, which can contain hundreds or as many as ten thousand galaxies. Such a concentration of matter can also distort the image of even more distant galaxies. Figure 4.3 shows what happens when a distant galaxy is directly behind a large galaxy cluster. The distant galaxy is distorted and stretched out in a series of arcs. It's similar to an Einstein ring, but because the lensing galaxy cluster is spread out, a true ring is not likely.

Such a dramatic series of arcs is pretty rare. After all, it requires very precise alignment. The galaxies that are more distant than the nearby galaxy cluster tend to be scattered all over the field of view. In this case, a far more subtle effect occurs—one that is too subtle to be seen by the human eye. What happens is that the images of all

Figure 4.3 The Starburst Arc galaxy shows four distinct arcs of a more distant galaxy, imaged by a closer galaxy cluster. The more distant galaxy is about 11 billion light-years away. The bright spots are foreground stars in the Milky Way galaxy. (Figure courtesy ESA/Hubble/NASA.)

of the very distant galaxies are just slightly distorted. Astronomers can then apply sophisticated algorithms that look at all of the distant galaxies and figure out how much mass would be required to be in the foreground galaxy cluster to account for the observed distortions.

And that is where the mystery arises. Astronomers look at the nearby galaxy cluster, counting up the galaxies and estimating how many stars there are in them. From that, they can work out the mass of the nearby galaxy cluster tied up in stars and hydrogen gas. They then compare that number to the amount of mass required to cause the distortion they observe. What they find is that there is more distortion than the visible matter can explain.

And that is yet another mystery. I've given three cosmological or astronomical mysteries where there is very clearly something wrong with our understanding of the cosmos. There are many possible explanations. Let's dig into it and see if we can rule out some of them or, even better, figure out which one is right.

Possible Solutions

OK. Where are we? We have three cosmic conundrums that need to be explained. It's possible that they have a single, common answer. It's also possible that the three mysteries are unrelated.

So what's going on? Well, that's the question, isn't it? It is often true that multiple disagreements between measurement and prediction originate from a common cause. Let's explore that possibility. If we can find no single solution, we can always back out and search for multiple ones.

For all three observations, the combination of observed matter and accepted physics disagrees with other measurements. How can we frame this tension in a simple way? I think the various possibilities can be most clearly illustrated by using Vera Rubin's galactic rotation measurements, so I'm going to start the discussion by boiling down the most crucial points for this particular conundrum.

The possible solutions of the galactic rotation mystery can perhaps be illustrated most easily with a faux equation. If stars are orbiting in more-or-less circular orbits, there must be a force that causes the stars to follow circular paths. And the only thing that is causing the stars to move in orbital paths is gravity. Accordingly, we can write:

Force(circular motion) = Force(due to gravity)

So this is a super simple idea. If the data and predictions of the orbital speed of stars around galaxies disagree, then we have only a few options. In short, the left side of the equation is wrong, the right side is, or the equal sign is wrong. It's as easy as that.

Being a bit more specific, the options are as follows:

Left side: Our ideas about circular motion are wrong. More generally, this is the same as saying that our ideas of inertia and Newton's laws of motion are wrong.

Equals: Maybe something more than gravity is what is causing stars to orbit. Essentially there is something missing in the equation.

Right side: We don't understand gravity as well as we think we do. There are two subcomponents of this. Option one is that Newton's law of gravity (and also Einstein's general relativity) is wrong. The second option is similar, but slightly different. Maybe the equations are just fine but, since gravity is caused by mass, maybe there's more mass than we can see.

Basically, those are the only options, and I will talk about each one in turn. Let's start with the hypothesis put forth by Fritz Zwicky nearly a century ago.

The Matter Solution

Zwicky coined the term "dark matter" in his 1933 paper in the journal *Helvetica Physica Acta*. This was one of his explanations for the discrepancy he observed between measurement and prediction of the motion of galaxies in the Coma cluster. Essentially, this hypothesis suggested that there was a large amount of additional matter that he could not see in his telescope.

This hypothesis was perfectly reasonable for the time. He could see galaxies, but he didn't have the instrumentation we have today. He couldn't see clouds of hydrogen gas that surround most galaxies. For that, you need radio telescopes, and that technology was still in his future. He also couldn't see infrared light or, for that matter, any other parts of the electromagnetic spectrum. With such a limited window on the cosmos, there was little hope of resolving the discrepancy, and the mystery lay dormant until the Rubin renaissance. But with Rubin's measurements telling a tale similar to Zwicky's, the scientific community set out to answer a question: "If the observed anomalies are due to overlooked or invisible matter, what kind is it, and how can we find it?"

So let's think it through. Dark matter could be some form of ordinary matter, much like the matter that makes up the visible stars and galaxies. Or it could be something exotic. In the same way that Roentgen's observation of a dim glow in a darkened room led to the discovery of X-rays, opening a window on previously unknown phenomena, perhaps Zwicky's and Rubin's observations could herald the discovery of something utterly unknown.

Of course, that second possibility is the less likely of the two. In the medical community, there is a phrase that is taught to medical students: "When you hear hoof steps, expect horses, not zebras." It reminds the student that when a patient shows up with a cough and a runny nose, it's more likely that they have the common cold than the possibility that they have some infectious disease found only in the jungles of Borneo.

In the same way, a priori it is far more reasonable to expect that dark matter is something ordinary—just dark. So scientists began thinking about what it could be. What form of matter could be highly prevalent, but invisible to optical telescopes?

Well, the most likely option is simply that galaxies and clusters of galaxies are enshrouded in clouds of hydrogen gas that have not yet coalesced into stars. While radio and infrared measurements of hydrogen gas were denied to Zwicky, by Rubin's time, the state of the art had advanced.

There are several different varieties of galactic hydrogen. There is neutral molecular hydrogen, consisting of two hydrogen atoms linked together (i.e., H_2). Then there is hydrogen that has been heated by nearby stars hot enough that the molecules break apart and individual hydrogen atoms are the result. Using modern techniques, astronomers have made reliable observations of all manner of interstellar and intergalactic hydrogen, and they find that there is a great deal of mass to be found in gas clouds. In many galaxies, about 10% of the mass is tied up in stars, while the remainder is tied up in gas clouds.

However, even when all of the mass from stars and gas clouds are combined, that's not enough. When one determines the mass of galaxies by looking at how fast they rotate, one finds that stars and gas make up only about 15% of the galaxy's mass. Clouds are not enough. Astronomers need to find more.

MACHOs

If the answer wasn't clouds of hydrogen, then what was plan B? The next obvious candidate is something similar to what we can see, but different—like stars that can't be seen. Perhaps the periphery of galaxies is full of black holes, dim stars, and similar, small objects. The periphery of galaxies is called the galaxy's halo, and consequently these hypothetical astronomical bodies are called MAssive Compact Halo Objects, or "MACHOs." Could dark matter be MACHOs? And how would you see them?

Searches for MACHOs rely on a physical phenomenon called "microlensing." It's the same basic idea as the gravitational lensing caused by galaxies and galaxy clusters that we talked about earlier. The difference with microlensing is that it is the gravitational lensing of a distant star by a closer star along the same line of sight.

Does this mean that we should see an Einstein ring around the nearby star? Well . . . yes and no. The physical phenomenon is the same, but the amount that a star will bend light that passes around it is much smaller than when a galaxy forms the lens. So no ring. Instead, what happens is that the light from the distant star brightens as the closer star passes in front of it. The general idea is shown in Figure 4.4.

This phenomenon provides an ideal method for detecting MACHOs. Although the term MACHO is more general, it is easier to visualize if we select a particular example to discuss.

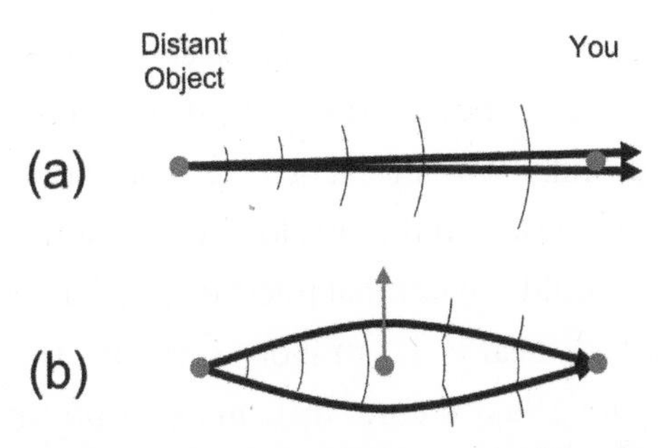

Figure 4.4 Microlensing occurs when an invisible object passes between an observer and a more distant object. Prior to the passage, a small amount of light is received by the observer; however, during the passage, more light is bent into the observer's eye, thus temporarily brightening the distant object.

So let's choose a black hole to guide our thinking. Black holes are the corpses of dead stars. They are very small, emit no light, and are very massive. This mass means that they exert a gravitational force larger than that exerted by our sun. In short, they are a perfect possible explanation for the dark matter mystery.

In order to find these black holes, researchers would look at a portion of the sky that is full of distant stars, and the brightness of those stars would be monitored. If a black hole passes in front of the distant star, the star will temporarily brighten and then dim back to its original brightness. Many stars are variable stars; this means that they vary in their brightness all on their own. In order to separate variable stars from microlensing events, it is crucial that when a distant star is seen to brighten that it does so identically in all wavelengths.

In 1992, a group of U.S. and Australian astronomers formed a collaboration with the name MACHO to search for . . . well . . . MACHOs. They employed a fifty-foot telescope located at the Mount Stromolo Observatory, near Canberra, Australia. For a distant star field, they chose two: the Large Magellanic Cloud (LMC) and the Milky Way's galactic core. For about seven years (1992–1999), they observed the sky, looking for some to brighten in the way that would suggest that microlensing had occurred.

They observed nearly 12 million stars in the LMC and 15 million in our galaxy's core. In January 2000, they released the results of 5.7 years of LMC observation, and they saw about fifteen microlensing events. From this result, they estimated that MACHOs accounted for about 20% of the dark matter necessary to explain the observed discrepancy in the Milky Way's rotation rate.

MACHO wasn't the only collaboration of astronomers studying microlensing. In addition, there was the Optical Gravitational Lensing Experiment (OGLE), which used a telescope located at the Las Campanas Observatory in Chile, which began in 1992, after the MACHO experiment began in Australia, and runs to the present day. (In recent years, the research focus has turned to exoplanet detection, but the ability to search for MACHOs remains.) Then there was the EROS collaboration (Expérience de Recherche d'Objets Sombres), which was located at the European Southern Observatory at La Silla, Chile. This collaboration studied both the LMC and the Small Magellanic Cloud (SMC).

Both EROS and OGLE found that the amount of dark matter in the form of MACHOs was quite small—smaller than that observed by the MACHO collaboration. When all of the data were analyzed, the researchers have definitively concluded that MACHOs might be a small fraction of dark matter, but they are definitely not the dominant type.

We must be slightly cautious. Technically, the groups have ruled out MACHOs in the mass range of 10^{-8} to 100 solar masses.

(10^{-8} solar masses is equivalent to about 30% of the mass of the moon.) Still, it would be extraordinary if there were MACHOs outside this mass range and not inside it, so most of the astronomical world has abandoned the MACHO hypothesis and moved on.

Hot or Cold?

If astronomers no longer believe that dark matter is some unusual configuration of ordinary matter, then where do we go after that? This is where things get tricky, because it might be that dark matter is of some form that we've never observed before. This allows theoretical physicists a great deal of freedom as they speculate about the nature of dark matter.

With the rejection of MACHOs, the scientific community has moved toward a model of dark matter that is more like a gas of some sort of unknown particles. The closest analog with ordinary matter would be if the entire universe were filled up with a bath of neutrons. Neutrons are not a candidate for dark matter because, although they have mass, neutrons also decay in about fifteen minutes. So they're not an option. But stable particles that are kind of like neutrons are quite possible.

On the other hand, this raises an important question. If dark matter is made of a gas of subatomic particles of unknown origin, there are two broad options. The particles could be heavy (compared to other subatomic particles) and relatively rare, or they could be light and plentiful. Are there any data that help physicists start to narrow down the properties of particulate dark matter?

Well, light particles tend to move quickly—often at a substantial fraction of the speed of light. Borrowing from the terminology of thermodynamics, a swarm of low-mass, fast-moving, particles is called hot dark matter. On the other hand, if the universe is instead filled with a bath of heavy and slow-moving particles, that's called cold dark matter.

While we've never directly observed dark matter, we believe that it would have been created at the Big Bang, just like ordinary matter. It would have started out highly energetic and lost some of its energy as the cosmos expanded. And it would have been present in the early universe and influenced the accretion of ordinary matter into stars and galaxies.

Luckily, the Hubble Space Telescope was launched in 1990, and it has spent over thirty years, peering ever deeper into space and, consequently, ever farther back in time. Researchers using this wondrous astronomical facility have been able to image galaxies from about 400 million years after the universe began. This is not long after the first galaxies began to form. So we have data that will help us understand whether dark matter is, broadly speaking, heavy or light.

We start by looking at the cosmic microwave background radiation (CMB), first discovered back in 1964. This radiation is the oldest snapshot ever observed of the conditions of the universe when it began. The CMB tells us about how uniform the universe was a scant 380,000 years after the universe began (and long before stars and galaxies formed). The important point is that the CMB is spectacularly uniform. Although small nonuniformities have been observed and are very important in guiding our understanding of the evolution of the cosmos, those uniformities really are small—in the ballpark of 0.001%. This means if you look anywhere in the early universe, the distribution of matter and energy was nearly identical.

Now we fast-forward to the time when galaxies are forming. Of course, for galaxies to form, clouds of hydrogen gas will have to have collapsed into pockets that are much denser than normal. It is this increased density that eventually resulted in stars.

How does hot or cold dark matter fit into the picture? If the primordial dark matter were hot, it would be moving very fast, which would allow it to escape any early clouds that began to form. Accordingly, hot dark matter predicts that galaxies would form by giant clouds the size of galaxy clusters coalescing and then those clouds would slowly break up and eventually make galaxies.

In contrast, if primordial dark matter were cold, it would be moving much more slowly. Because of its low speed, it would tend to stick around and reinforce the collapse of hydrogen clouds. The consequence of cold dark matter is that small clumps of gas would form first, creating not galaxies, but what one might call "protogalaxies," with masses of thousands, tens of thousands, or even hundreds of thousands times the mass of the Sun. This is in stark contrast with the mass of, for example, the Milky Way, which has a mass of hundreds of billions or possibly even a trillion solar masses.

These protogalaxies would then slowly coalesce and galaxies would grow larger and larger, becoming the monstrously large leviathans that we see today. We see that cold and hot dark matter give very different predictions of the evolution of galaxies and galaxy clusters. So what does the data favor?

A dramatic drumroll and the answer is . . . it appears that primordial dark matter was slowly moving and, by extension, dense. This seems to rule out, for example, the idea that neutrinos formed during the Big Bang are dark matter. The known neutrinos are extremely light and, while there are many still around from when the universe was created, they just don't have the right properties. There must be another explanation.

Yet, when we look to the standard model discussed in Chapter 2, we find no credible candidates. In order for dark matter to be cosmologically important, it has to have existed since the first moments of the universe; thus it must have existed for 13.8 billion years. Thus, dark matter must be stable on cosmic timescales. Individual dark matter particles must be heavy—perhaps a hundred times heavier than a proton or neutron, or maybe even heavier than that. It has to be electrically neutral, and it cannot contain within it quarks or any other known denizens of the standard model. If it did, then cosmic rays and highly energetic gamma rays from distant and ancient galaxies would interact with it on its journey to be detected here on Earth. And we don't see any evidence for that. Thus, scientists have concluded that if the explanation of the phenomena studied by Zwicky and Rubin is indeed dark matter, then dark matter must be something entirely new and a particle that doesn't exist in our current theories.

WIMPs and Other Heavy Ideas

Without a discovery of dark matter, scientists don't have an agreed-upon name for it. Instead, a placeholder name was devised. In keeping with the irreverent, tongue-in-cheek, and just downright silly, naming culture of particle physics, the hypothetical dark matter particle is called a WIMP, short for Weakly Interacting Massive Particle. And, of course, wimps aren't macho and vice versa, so it seemed that particle physics and English were in agreement. For the record, the term WIMP predated MACHOs.

What are WIMPs? Well, there is no unique definition for them, but broadly we can think of WIMPs as massive, stable, subatomic particles that interact with gravity. However, since gravity is so incredibly weak, we'll never detect them if they only interact gravitationally. Thus, physicists also postulate that WIMPs also interact with some other force that also interacts with ordinary matter. That force could be the weak nuclear force (although this is now disfavored) or any other force that is no stronger than the weak force—and maybe much weaker—but also much stronger than gravity.

This last requirement—that there must be some sort of weak or super weak force that dark matter interacts with—isn't, strictly speaking, required from the astronomical measurements. According to astronomy, only gravity matters. So the requirement that another force comes into play is simply to imagine how we might detect and study WIMPs.

When I was just a wee lad—fresh out of graduate school—there was some hope that dark matter also interacted via the well-known weak nuclear force. If dark matter were formed via thermal production, it must have a specific self-annihilation probability, meaning dark matter and anti-dark matter would have to annihilate just like ordinary matter and antimatter do. And, under these conditions and using the electroweak force to mediate that interaction, theorists predicted that the mass of dark matter would be about 100 times heavier than a proton. That is great news for particle physics researchers, as we have constructed particle accelerators that could create such a particle. Furthermore, researchers looking for dark matter flowing through the solar system, "in the wild" so to speak, had a prayer of detecting it. We'll talk about experimental studies that are searching for dark matter shortly.

If the weak nuclear force doesn't interact with WIMPs, then there must be some new and currently unknown force that does. We don't know anything about that force and won't until we discover dark matter and then work backward to understand this unknown force's properties.

What ideas have been put forth as candidates for dark matter? Well, the most popular suggestion from the 1990s through the late 2010s originates from the theory of supersymmetry that we discussed in the section describing superstrings in Chapter 3. Briefly, a theory is supersymmetric if the equations that describe nature have the property that we can swap all of the fermion (half-integer spin) and boson (integer spin) particles and not change the equation.

The simplest way one would write a supersymmetric version of the standard model is to use the equations that we currently understand and add new terms. For every fermion term in the known equations, add a new boson term, and vice versa. One consequence of adding these extra terms is that they predict that every known subatomic particle of the standard model has a supersymmetric cousin of the opposite form. For example, the electron is a fermion, with a spin of ½. Supersymmetry then predicts a new particle, called a "selectron," which is a boson with spin 1. Similarly, the standard model photon is a boson, and a supersymmetric theory predicts that a corresponding supersymmetric fermion called a "photino" should exist.

In general, in supersymmetry theory, the bosonic counterparts of the familiar fermions all have the same name with an "s" in front of it, giving squarks, smuons, sneutrinos, and so on. Conversely, the supersymmetric fermion components of the familiar bosons keep the boson name and add "ino" on the end, sometimes with a little spelling change to make the names friendlier, like Wino, Zino, gluino, and so on.

Now because we've never seen any of these supersymmetric cousins, theorists postulate that they are unstable, decaying to the particles of the standard model. However, supersymmetry theory generally predicts that the lightest of the supersymmetric particles (a.k.a. the LSP) is stable on cosmic timescales. And, because we haven't found it yet, the LSP must interact very rarely, and it must be heavy. In short, the LSP is a perfect candidate for dark matter, which is one reason why supersymmetry was so popular in the theoretical community at the turn of the last century. Thus far, no LSPs have been observed—indeed, no indication that supersymmetry is true has been observed either.

Another possible dark matter candidate is a hypothetical kind of neutrino called a sterile neutrino. The neutrino is called "sterile," because it is postulated to be much like a neutrino, but without the ability to interact via the weak nuclear force. Furthermore, some theories of neutrinos claim that they don't get their mass from interacting with the Higgs field, but rather in another way, called the seesaw mechanism. In that theory, the mass of a neutrino and sterile neutrinos are coupled in such a way that if one goes up, the other goes down (hence the term "seesaw"). And, because neutrinos have such a small mass, then sterile neutrinos should be very massive. A very massive, stable, electrically neutral particle is the very definition of dark matter, so sterile neutrinos should be added to the stable as dark matter candidates.

While I've mentioned two possible WIMP candidates, many others have been proposed. The only way we'll know which (if any) of the proposed dark matter candidates is right will be to observe them. And that leads us to an important question. What attempts have been made to detect dark matter particles, beyond the mysteries with which I started this chapter?

Observation Strategies

Because dark matter (if it exists) interacts only via gravity and possibly some other very weak force, it is very difficult to detect. There are three methodologies that scientists are using to try to find dark matter. These methods are called indirect detection, direct detection, and laboratory creation.

Indirect detection simply looks for the signature of dark matter somewhere "out there" in the universe. The methodology is simple in principle. If dark matter exists and there is an antimatter version of dark matter, then the two should be able to annihilate, just as matter and antimatter do. Accordingly, scientists simply look in places where a lot of dark matter appears to congregate and look to see annihilation products originating there.

The actual hypothesized dark matter annihilation process is a bit complicated, but it is thought that the two heavy particles will disappear and release high-energy gamma rays. Gamma rays are electrically neutral, so they race across the cosmos, undeflected by the magnetic fields that permeate interstellar and intergalactic space.

Zwicky and Rubin's measurements suggest that dark matter is densest in the center of galaxies or galaxy clusters. Also there is a class of small galaxies in which dark matter is far more prevalent than usual. In all of these cases, what scientists do is look toward these theorized locations of high-density dark matter and look for high-energy gamma rays coming from them.

There are a number of gamma ray telescopes that search for high-energy gamma rays. These detectors are generally not solely tasked to do indirect studies of dark matter; instead, they were built as general-purpose observatories dedicated to gamma ray astronomy. The ability to search for dark matter is just a welcome spin-off. Some of the noteworthy detectors are the orbiting Large Area Telescope aboard the Fermi Gamma-ray Space Telescope (Fermi-LAT); VERITAS (Very Energetic Radiation Imaging Telescope Array System), located at the Fred Lawrence Whipple Observatory in southern Arizona; and HESS (High Energy Stereoscopic System, located in Namibia, Africa).

To date, none of these facilities have announced a universally accepted signal of high-energy gamma rays, although Fermi-LAT has reported an excess of gamma rays coming from the center of the Milky Way. Many papers have been written, explaining the excess as coming from dark matter annihilation, although as astronomers have improved their understanding of such things as pulsars, the dark matter explanation is becoming less persuasive.

This points to a very important consideration when one is doing indirect dark matter searches. Basically, the universe is a complicated, mysterious, poorly understood, and violent place. Stars explode in spectacular supernovae. Black holes are known to be voracious eaters, but they are sloppy ones, and they shoot matter out into space at high speeds. Pulsars—stars the size of a city that spin many times a second—host strong magnetic fields that can play havoc on the material surrounding them. All of these (and more) are candidates for creating more gamma rays than their more sedate astronomical neighbors, and to claim you've discovered dark matter by indirect means, you have to have a deep understanding of these crazy denizens of the cosmic zoo.

There are a number of ways in which scientists perform indirect studies of dark matter, but perhaps one more is worth mentioning here. One possible location where there might be a lot of dark matter is the center of the Sun. The basic idea is that dark matter in our galaxy will occasionally pass through the Sun. Since the Sun is big and dense, dark matter might interact a little more in the solar interior and lose enough energy to be trapped. If that happens, slowly but surely dark matter and its antimatter equivalent will accumulate at the core of the Sun. And, if a dark matter/antimatter pair meet and annihilate, we might have a chance of seeing that here on Earth.

Of course, an annihilation producing gamma rays wouldn't work. After all, the Sun is largely opaque to gamma rays, and it takes many thousands of years for them to make it from the core to the surface. However, there are models of dark matter/antimatter annihilation that result in neutrinos being produced, and neutrinos can punch through the Sun without any problem.

Accordingly, scientists can study the Sun, looking for very high-energy neutrinos originating from it. In the normal solar processes, the neutrino energies are quite modest. However, dark matter/antimatter annihilation would result in super high-energy neutrinos, so seeing an excess would be a possible signature for the presence of dark matter in the Sun.

Astronomers use giant neutrino detectors, like those mentioned in the previous sections (like the IMB and Kamiokande experiments used to search for proton decay, but bigger and more modern) to search for an excess of high-energy neutrinos, without luck. Or, more accurately, not enough neutrinos have been observed to rule in or out any theory at all. Indirect dark matter detection is a very difficult business.

So, if indirect detection is hard, what about direct detection? How does that work?

If dark matter is real, then it doesn't just exist in the outskirts of galaxies, or in the depths of extragalactic space. It exists everywhere, in varying densities. It even exists within our own solar system. In fact, it is thought that over the course of the Sun's 4.5 billion-year lifetime, that the Sun has slowly concentrated dark matter in the vicinity of the Earth by about 10,000 times greater than the average dark matter density in interstellar space in the vicinity of our solar system. On Earth, the dark matter density is thought to be equivalent to about 6,000 hydrogen atoms per cubic centimeter.

That sounds like a lot, but when one scales it to the size of the Earth, that means that there is about 11,000 kg of dark matter in the Earth at any one moment, or about eight automobiles. On the other hand, the mass of the Earth is about 6×10^{24} kg. So dark matter inside the Earth is very diffuse. (If it exists at all.)

But it's even more diffuse in interstellar space, about 10,000 times less dense. Interstellar dark matter corresponds to about a kilogram's worth in the volume of the Earth and that's super diffuse.

And it's the interstellar form of dark matter that is important for discovering in Earth-based detectors. That's because the Earth moves through the interstellar dark matter with a speed of about 270 kilometers per second. And that speed is an important part of the strategy for finding dark matter. So how does that work?

Remember that WIMPs are defined to be subatomic dark matter particles that experience both gravity *and* some other weak force that might be weaker than the weak nuclear force, but is way stronger than gravity. We don't know anything about that second weak force, but we assume it exists.

And, if that unknown weak force exists, the dark matter might interact in a detector. If dark matter only interacts via gravity, it is unlikely that we'll ever directly detect it here on Earth.

OK, so those are the two unknown parameters of dark matter. The first is the mass of individual dark matter particles, and the second is how strong the force is whereby it interacts with ordinary matter. Any experiment searching for dark matter must be able to winnow out those two parameters from the data.

So how do dark matter detectors directly searching for it work? Well, there have been literally dozens of different experiments, using different technologies, but they all have some commonalities. The idea is that the interstellar dark matter wind will blow through the detector and, very rarely, a dark matter particle will collide with an atom in the detector. That atom will move, and we'll use the moving atom to tell us that dark matter interacted.

There are some enormous technical issues. For instance, atoms are always moving from ordinary heat. That's what atoms do. They jiggle around, bouncing into one another and generating general atomic havoc. So, to tame the motion of atoms, dark matter detectors are often cooled to very low temperatures, which can range from the 162 K (–108 °C, –163 °F) temperature of liquid xenon, to sometimes as low as a fraction of a degree above absolute zero. At such low temperatures, when an atom is seen to move, it could be due to the passage of dark matter particles.

Of course, not only dark matter particles will bump into atoms. Cosmic rays from space, particles emitted by radioactive decay in the surroundings, and even radioactivity that occurs in the detector can all bump into atoms. So scientists work hard to shield their detectors from these unwanted interactions. They often place the detector a kilometer or more underground to shield it from cosmic rays. They surround the detector with radiation-sensing detectors that will tell them if a radioactive decay occurred outside their detector. And they build the detectors out of materials that have very low intrinsic radioactivity.

After all of those precautions are achieved, the researchers wait, hoping to see evidence of that wispy wind of dark matter that flows through their apparatus.

So what have scientists seen? Well, by and large, nothing. There have been a few scientists who claim to see a signal that is consistent with the dark matter wind, but those measurements have not been confirmed by other experiments and, often, the tentative observations have been ruled out by more powerful and sensitive detectors.

The field is constantly evolving but, as of 2022, the best technology for directly searching for dark matter wind here on Earth involves large vats of liquid xenon, containing hundreds or even up to a thousand kilograms of the material. None of these experiments have seen any evidence for dark matter.

And it's not for a lack of trying. Over the last decade alone, detectors have become ever more sensitive, with the newer ones having sensitivity over a million times better than earlier ones. Researchers still have a way to go, but it won't be too long before the dark matter detectors will begin to see the bath of solar and atmospheric neutrinos that constantly passes through the Earth. At that level of sensitivity, continued searches for dark matter will become increasingly difficult, as the desired signal will be drowned out by the cacophony of neutrinos.

The bottom line is that direct detection of dark matter has not yet been successful, and it may never be.

There is one final method whereby researchers might be able to discover dark matter, and this one is my personal favorite. Rather than looking for dark matter "in the wild," so to speak, researchers are trying to create it in laboratories using large particle accelerators.

Scientists like me have a long history of using particle accelerators to create particles that don't exist for very long in nature. Using the magic of Einstein's $E = mc^2$, we accelerate particles to near the speed of light and slam them together. The energy of the collision is converted into particles. That's how we found the heavier quarks and the Higgs boson. As long as dark matter interacts with a force that isn't too much weaker than the weak nuclear force, we have a prayer of success.

There have been many (and I do mean many) theoretical models proposed that have predicted different forms of dark matter. It is not possible in a short treatment of the topic to cover them all.

The most popular theories with consequences for dark matter are ones that include supersymmetry. Most supersymmetric models include a lightest supersymmetric particle that is an ideal candidate for dark matter. However, just as is the case with the direct detection methodology, no evidence for dark matter has ever been found in a particle physics collider.

Scientists have pursued the three different methodologies for finding dark matter: direct, indirect, and lab created, for many decades and have come up unsuccessful again and again. Does that mean that the dark matter hypothesis has been falsified? No, not really. It is always possible that dark matter particles are heavier or interact more weakly than all of the different methods can detect. So scientists will continue to search. By the time you read this, dark matter might have been discovered. Or not.

Alternatives to Dark Matter

Given the long and unsuccessful history in searching for dark matter, we should consider some of the other alternatives. As I described in the section discussing Vera Rubin, it is possible that dark matter doesn't actually exist. Instead, the other big possibility is that we don't fully understand the laws of motion or how gravity works.

In this section, I'll talk about those two possibilities. While they seem to be different, they are more intertwined than appears at first blush. Hopefully it will all be clear.

The first attempt to come up with a non–dark matter explanation for the glut of astronomical mysteries that became apparent in the 1970s was made by Israeli physicist Mordehai Milgrom in 1982. He came up with an explanation that approached the problem by assuming that Newton's second law of motion was incomplete.

Newton's second law states that the acceleration an object experiences is directly proportional to the force applied to it. (If you ever took a physics class and remember any of it, this is the equation $F = ma$.) Increase a force by 50%, and the acceleration will go up by the same amount.

Milgrom decided to use Rubin's measurements of the rotation rates of galaxies to motivate his thinking. As a reminder, she showed that at large distances from the galactic center that stars orbited at roughly constant speed, while Newton's equations predicted that the orbital speed would get slower and slower as one looks at stars far from the center. What Milgrom wanted to do was to generate equations that would predict a stellar speed that was constant, independent of radius (at large orbital radii), while keeping the behavior predicted by Newton at small orbital radii.

In the case of galactic rotation, the force (and consequently acceleration) is larger near the center and weaker at the periphery. So he proposed that for large accelerations, Newton's law still held. However, for smaller accelerations, he suggested a new equation.

The acceleration due to gravity depends on one over the radius squared ($1/r^2$), while the acceleration required to move in a circle depends on one over the radius ($1/r$). Since the two accelerations are on opposite sides of the equal sign (of the motion equals gravity equation), and he wanted to get rid of any radial dependence, he simply modified Newton's second law from being the familiar version, which says force is proportional to acceleration ($F \propto a$), to a new equation that says that force is proportional to acceleration squared ($F \propto a^2$).

This choice put a one over radius squared on both sides of the equation, which then cancelled out. The result was that he got the result he wanted: At large distances from the center of galaxies, stars moved with constant velocity.

Of course, he got that result because he engineered the equations to get the desired outcome. That's not necessarily a bad thing; after all, looking at data and devising equations that agree with it is a time-honored technique for trying to understand the natural world. However, his equations didn't come out of any underlying principles or theory, and that's an important thing to keep in mind.

Milgrom's approach is known under the ill-defined term of MOdified Newtonian Dynamics (or MOND, for short). One has to be cautious when a person uses the term MOND, as it is often used as a blanket term that is roughly analogous to "not dark matter." The only way you can understand exactly what they mean requires that you pay close attention to context. It's an unsatisfying state of affairs, but there you have it.

When we talk specifically about Milgrom's MOND, we have to be a bit careful. There is a general form, which applies to the motion of all objects, under all forces and all circumstances, and then there is his MOND when applied specifically to the force of gravity and circular motion.

Milgrom's original version of MOND said the following. There is a specific acceleration that separates how objects move. Above that acceleration, Newton's familiar second law works perfectly—force is proportional to acceleration. Below that acceleration, Newton's second law no longer works. Instead, force is proportional to acceleration squared. And he was fuzzy about transition from one behavior to the other.

When Milgrom applied his formula to Rubin's galactic rotation data, he found that the acceleration that separated Newtonian behavior from his modified version was about one hundred-billionth (10^{-11}) the acceleration due to gravity that we experience on the surface of Earth. This is good news to all introductory physics students, who routinely test Newton's second law in their laboratory as part of their studies. The students always find that Newton was right.

However, the students never test Newton's laws for the tiny accelerations that Milgrom is modeling, meaning that his hypothesis remains untested in the laboratory.

The fact that Milgrom's theory accurately predicts the galactic rotation curves is not a validation of the theory. After all, his modification was added precisely to do just that. In order for his theory to be well regarded, it needs to be validated with other measurements, like the motion of galaxy clusters and the gravitational lensing that was described earlier. And we'll get to that in a little bit.

But first, there are some significant issues with Milgrom's initial theory. For instance, the theory was strictly nonrelativistic, which is a problem if it is to be taken as a serious proposal. That problem was overcome in 2004 when Jacob Bekenstein devised the first fully relativistic version of MOND, called TeVeS, short for "Tensor-Vector-Scalar" gravity. This theory is too complicated to describe fully here, but, being fully relativistic, it can make predictions about gravitational lensing, unlike the first version of MOND.

While the existence of dark matter or modifications of the laws of motion are two possible solutions to the cosmic anomalies mentioned at the beginning of this chapter, there exists yet a third possibility. Perhaps we simply don't understand gravity properly. Perhaps—and this sounds heretical to many people—perhaps Einstein's theory of general relativity isn't completely correct.

Many alternative theories of gravity have been proposed since the creation of general relativity. Since the 1970s, an increasing number of them have been proposed as alternatives to the dark matter hypothesis. All of the theories are substantially more complex than general relativity and are well beyond the scope of this book. Suffice it to say that they exist. Additional information is available in the Suggested Reading.

Status

So what's the answer? Is dark matter real or not? You hear all the time that dark matter is five times more prevalent than ordinary matter. Are physicists pulling your leg, or what?

As I hope you're now convinced, the honest answer is "we just don't know." While most physicists and astronomers find the dark matter situation to be more persuasive, the MOND alternative is still viable. And a small number of passionate researchers are convinced that some version of MOND or alternative gravity is the answer.

Each camp points to the successes of their model and the deficiencies of the other. I think the best we can say is that the jury is still out. I think the situation is well summarized by the illustration in Figure 4.5. At the root of the tree are the various astronomical observations that cannot be explained by Newton's well-tested laws of motion and theory of gravity, and the visible matter in the universe. As one goes up the trunk of the tree, one encounters the first split into two boughs; one of the options is that missing matter explains the discrepancy, while the other selects the possibility that our understanding of inertia or gravity is wrong.

The boughs then split again into smaller branches, like the split between alternate theories of motion or gravity, or the split into dark matter being unseen ordinary matter or something exotic. Subsequent branches result in ever more specific possibilities until one reaches the crown of the tree, where individual leaves represent specific theories. Over the past few decades, researchers have pruned the tree here and there, lopping off twigs that are no longer tenable (e.g., MACHOs) and sometimes discovering new branches,

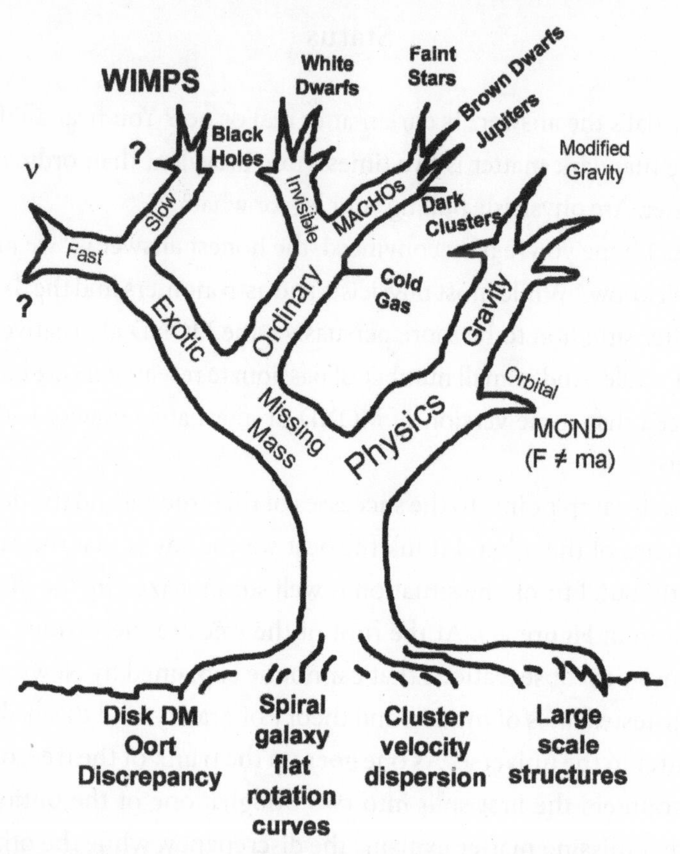

Figure 4.5 The conundrum of dark matter is unsolved, a situation which is well represented by this metaphor. At the root of the problem are a number of unexplained observations. There are several "big-picture" possible explanations, like unobserved mass (i.e., dark matter) or an incomplete understanding of the laws of physics. Each big-picture explanation has several possible subcategories, and the final answer will be just a twig on the tree. (Figure courtesy Stacy McGaugh.)

as additional ideas are proposed. It could be a very long time until the mystery is resolved.

And this brings us back to the main question of this book, which is progress toward a theory of everything and the ever-important

Figure 3.8. If the dark matter hypothesis is correct, then dark matter will have to fit in somewhere over on the right-hand side of the image, possibly with an extension of known forces and particles. Alternatively, if dark matter exists and the WIMP hypothesis is validated, then this means that there is probably another subatomic force that is weaker than the weak nuclear force. If that's the case, we'll have to add bubbles over in the region of the figure where the standard model now sits.

If, on the other hand, the answer to the dark matter mystery isn't actually dark matter, but rather a modification of gravity, then Figure 3.8 will have to be modified over on the left-hand side. We don't know if the relevant modification would be quantum or classical in nature, but it is almost certainly classical. That either means that general relativity will have to be swapped out and an improved theory of gravity swapped in, or perhaps the new theory will require a new bubble and some new connecting lines.

The key point is that what we might colloquially call the "dark matter mysteries" are very mysterious. Dark matter, if it is real, is five times more prevalent than ordinary matter, meaning that the structures on the right-hand side of Figure 3.8 are only a small part of the story, and some new dark matter bubble is really dominant. And, of course, maybe instead we'll have to reconfigure the left-hand side of the figure. Until this is figured out, we can't really take the dashed lines of Figure 3.8 seriously. Indeed, when this mystery *is* figured out, Figure 3.8 might look quite different.

In this chapter, we've talked about a proposed form of matter that is more prevalent than ordinary matter, but we're not done. Because of Einstein's famous equation, $E = mc^2$, we know that we can't separate energy and matter. And, when we take this into consideration, we find that when we combine dark and ordinary matter together, it only accounts for less than a third of the energy and matter of the universe. In short, there's a much bigger mystery to be solved and that's our next topic.

CHAPTER 5

DARK ENERGY

Fourteen billion years ago, the universe began expanding at a furious pace. We call this period of expansion the Big Bang. And it's still going on—after all, the universe is still expanding.

On astronomical size scales, the force that dominates is gravity and, as we all know, gravity is an attractive force—what goes up must come down and all that. These two facts—that the Big Bang created all of the matter in the universe and that gravity is an attractive force—allow scientists to ask and answer many questions, including weighing the universe and determining its future.

So how is that done? Well, it relies on the fact that at the moment just after the universe began, that the expansion rate of the universe was as fast as it gets. Then, for the next 14 billion years, gravity has been slowing down that expansion. We can then look into the universe, from near distances to far away, and see how the expansion speed has changed over time.

From what we know about gravity, it would seem that there are three possible outcomes of that study. The first is that there is a lot of mass in the universe. In that scenario, the strength of gravity overwhelms the expansion caused by the Big Bang and the universe slows to zero expansion and then collapses back on itself in a "Big Crunch." In familiar terms, it's similar to throwing a baseball

Einstein's Unfinished Dream. Don Lincoln, Oxford University Press. © Oxford University Press 2023.
DOI: 10.1093/oso/9780197638033.003.0005

upward at speeds accessible to the human arm. The ball goes up and falls back down. That's scenario one.

The second possible scenario is that there is a relatively small mass in the universe. In this scenario, gravity still slows the expansion, but gravity is too weak to stop the expansion. The universe will continue to grow forever, slowing over the eons, but never stopping. This is similar to shooting a baseball upward with some sort of super cannon that shoots the ball so fast that the ball breaks free of the Earth's gravity and soars off into interstellar space. (And, of course, we're ignoring the gravitational effects of the Sun and other planets.) That's scenario number two.

The third scenario is what one might call the Goldilocks' option—not too much, nor too little, but just right. In this situation, the expansion of the universe is slowed by gravity and sometime in the infinitely far future, the expansion stops and there is no crunch. In our ball metaphor, this is when the super cannon shoots the ball upward and, in the far, far future, it eventually gets ridiculously far away from Earth and stops. That's scenario three.

Depending on which scenario astronomers see, that allows them to determine the mass of the universe—is it the precise mass at which the expansion eventually stops without a follow-on crunch? Or is it bigger or smaller? And, of course, this also predicts the future evolution of the universe.

Astronomers have actually done this study and know the answer. So what did they see? There are only three options: Was it one, two, or three? Let's talk about this particular measurement.

The Future of the Universe

In the 1990s, astronomers were looking toward the heavens, trying to understand the expansion history of the cosmos. This is actually a tricky question to answer, requiring that they measure both the distance and velocity of objects that are literally billions of light-years away.

Measuring the velocity of astronomical objects isn't so hard. Back in 1912, American astronomer Vesto Slipher studied many galaxies, looking at the spectrum of light emitted by them. Galaxies are made of stars, and stars are big balls of mainly hydrogen gas undergoing nuclear fusion. These stars emitted light at many wavelengths, but there are certain wavelengths that are universal. What he found was that, while all of the galaxies he studied emitted what looked like those universal wavelengths, for most galaxies the wavelengths weren't in the same place as they are for nearby stars or when the same spectral lines were created here on Earth.

By 1925, Slipher had convinced the astronomical community that what he was seeing was an example of the Doppler shift. The Doppler shift is a universal phenomenon of waves, and light is a wave. This shift says that if an object is moving toward you, it will appear bluer—which is to say a shorter wavelength—than if that same object were stationary with respect to you. Conversely, objects moving away from you will appear redder. In order to see shifts in light spectra, the objects have to be moving very fast and you need rather precise instruments.

However, the Doppler effect is probably something you've encountered in your day-to-day life, specifically the audio version.

Sound is also a wave, and if you have ever heard a train whistle as the train passed by you, or watched a high-speed automobile race, you've heard the pitch change as the train or car passes by. As the vehicle approached you, it had a higher pitch than after it passed. That's the Doppler shift.

Slipher realized that the vast majority of galaxies were moving away from the Earth and, depending on how much of a shift he saw for each galaxy, he knew how fast they were moving.

Finding the distance to astronomical objects is much harder. For that, we need to talk about the work of Henrietta Swan Leavitt. Her advance in measuring interstellar, and even intergalactic, distances is one of the most important contributions to modern astronomy.

Determining the location of any star on the sphere of the heavens is easy to do. Astronomers have long been able to measure angles between stars and to give each star's latitude and longitude on the sky. However, distances are much trickier. If a star appears to be bright in the night sky, is it truly a bright star located a great distance from Earth? Or is it a dim star that is nearby?

For nearby stars, geometry is used. Astronomers pick a star out of the sky and measure its position in the sky on one night. Then, six months later, they measure the position of the same star. If it's close enough, then it will appear to have moved compared to more distant stars. Researchers can then use the two angles measured half a year apart and combine it with the 186 million miles that is twice the radius of the Earth's orbit. That sets up a triangle, and it is easy to work out the star's distance. The basics of this method are shown in Figure 5.1, and it was successfully performed for the first time in 1838 by German polymath Friedrich Bessel using the star 61 Cygni, located about 11.4 light-years away.

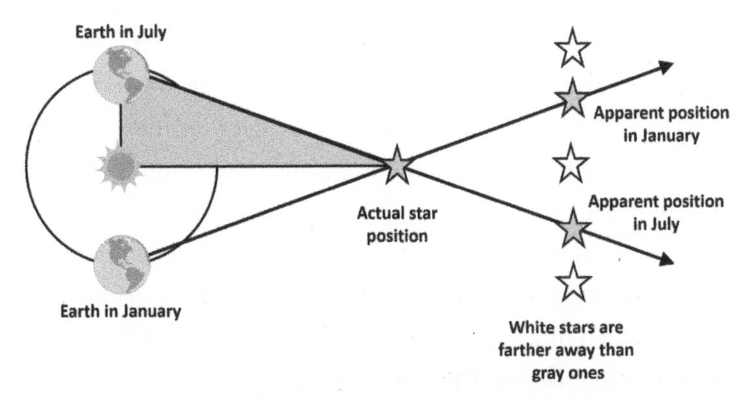

Figure 5.1 When a nearby star is viewed six months apart, it appears to move against a field of more distant stars.

This geometric method only works for nearby stars. Before Henrietta Leavitt, astronomers had no method to determine how far away more distant stars were.

Leavitt worked in the laboratory of Edward Pickering, director of the Harvard College Observatory. She was one of many female "computers" he employed. He tasked her with cataloging the brightness of stars that his staff had photographed using a telescope located in Peru.

Among the many stars she catalogued, she studied what are called Cepheid variable stars, which would vary in brightness over the course of days, weeks, or even months. Leavitt was interested in finding out if there was a relationship between the period of oscillation of the stars' brightness and their baseline brightness, but she encountered the same problem that had befuddled other astronomers at the time: not knowing the star's distance.

So she did a clever thing. She looked at Cepheid variables located in the Magellanic Clouds that are observed in the Earth's Southern

Hemisphere. The Magellanic Clouds are dwarf galaxies that orbit the Milky Way, although this was not known at the time. However, she reasoned that, whatever the distance between the Earth and the Magellanic Clouds, at least the distance was roughly the same for all of them.

What Leavitt found was that there was indeed a correlation between brightness and oscillation period. Brighter Cepheid variable stars oscillated more slowly than dimmer ones. She published her first results in 1908, with a follow-on paper in 1912.

She still didn't know how far away the stars were; however, in 1913 astronomer Ejnar Hertzsprung used the geometric method to measure the distance to a handful of nearby Cepheid variable stars. By combining Hertzsprung and Leavitt's work, astronomers could now use the oscillation period of Cepheid variables to determine their absolute brightness. They could then measure the star's apparent brightness in their telescopes and use the ratio of apparent to absolute brightness to determine the star's distance.

Leavitt's work was invaluable to astronomy, allowing researchers to measure distances as far away as a few million light-years. These distance measurements were used by Edwin Hubble to determine that there were other galaxies in the universe and also that the universe is expanding. This buttressed the idea of the Big Bang.

Hubble's key observation was a relationship between the distance between galaxies and the Earth, and the velocity at which they are moving away from us. This makes sense if the universe is growing at a constant rate.

Let me give you an example. Say that over some timeframe the universe doubled in size. Under this scenario, a galaxy that is 10 million light-years away will move to a distance 20 million light-years from Earth. In that time, the galaxy moved 10 million light-years.

In contrast, if distances in the universe double in size, then a second galaxy that is initially 20 million light-years away will have moved to a distance of 40 million light-years. In the same amount of time that the nearby galaxy moved 10 million light-years, the more distant galaxy moved 20 million light-years.

One can continue this trend to ever more distant galaxies, and one finds that the farther away a galaxy is from Earth, the faster that it is moving if the universe is expanding by some fixed factor. Hubble worked out a relationship between distance and velocity. His first measurement wasn't very accurate, but we now can say that for every 3.26 million light-years distance between the Earth and a galaxy, the galaxy is moving about an additional seventy kilometers per second. (The 3.26 number comes from using a somewhat obsolete distance measurement called a parsec. One parsec is 3.26 light-years.)

This basic relationship (speed = constant × distance) and where it eventually falls apart will play a crucial part in our story.

While Leavitt's work was crucial in understanding that the universe is expanding, in the 1930s, it was only useful out to a distance of a few million light-years, encompassing a few tens of galaxies. The universe is over 10 billion years old, so astronomers needed another trick to figure out distances. And, for that, they turned to one of the most cataclysmic events in the universe—the catastrophic explosion of a star, called a supernova.

It turns out that there are several different ways in which stars can explode, but a particular one is especially useful to astronomers, and it is called a Type 1a supernova.

Type 1a supernovae are formed in binary star systems. At least one of the stars is a white dwarf, which is a star that has used up all of its nuclear fuel and is no longer undergoing nuclear fusion. The star consists of mostly carbon and oxygen. Because of gravity, the white dwarf slowly siphons off mass from the other star, getting heavier and heavier. When the mass of the white dwarf reaches 1.44 times the mass of the Sun, the mass is now high enough to reignite fusion, converting the carbon and oxygen into heavier elements. This occurs very quickly and generates a tremendous amount of heat. The pressure from heating overcomes the force of gravity, and the white dwarf explodes. That's a Type 1a supernova.

Because the mechanism is so specific, Type 1a supernova all have a similar initial brightness, or at least astronomers can calculate how bright the star is when it explodes. They can then compare that intrinsic brightness to the brightness they see in their telescopes and determine just exactly how far away the star is.

Supernovae will temporarily outshine the galaxy in which it is housed. Thus, astronomers can use the information from the supernova to work out the distance of the parent galaxy—and those can be very far away—billions of light-years.

So it was in the late 1990s when two astronomical collaborations used Type 1a supernovae to measure how far away distant galaxies are and then used the Doppler shift to determine the galaxies' velocities. The collaborations were called the High-z Supernova Search Team and the Supernova Cosmology Project. ("z" is a technical variable that is related to cosmic distances and "high-z" means far away.) The two groups were studying galaxies that hosted supernovae from neighborly cosmic distances to ones as far away as about 7 billion light-years.

So what were they looking for? They were trying to understand how the expansion of the universe changed over time. If there was no change, then they'd expect to essentially reproduce Edwin Hubble's studies of the late 1920s, although at distances about a thousand times more distant. However, the expectation was that when the universe began, it was expanding faster than it is now. Then, over the cosmic eons, the universe would slow down. Thus, they expected that the galaxies very far away (and thus being viewed very far back in time) would be moving faster than ones much closer to Earth. (And, to do this, they needed to account for Hubble's observation that distant galaxies already move faster away from the Earth than nearby ones.) Essentially, the collaborations were looking for deviations from projecting Hubble's observations of nearby galaxies.

What I just wrote is conceptually what astronomers did. Technically, they did something that is different, but equivalent. They looked at the brightness of the supernovae as seen by their telescopes. The brightness is expressed in terms of a variable called stellar magnitude, with dimmer stars, galaxies, and so on having a larger magnitude. If you are an astronomy buff, the brightness of the supernovae they investigated was in the range of an apparent magnitude of 20 and 25, which is between 100 billion and 10 trillion times fainter than the star Vega. If Type 1a supernovae have a common intrinsic brightness, then the observed brightness tells astronomers the distance between them and the Earth.

While apparent magnitude stands in for a measurement of distance, for velocity, what researchers used is a variable called "z." As we will see, this variable is trickier to understand. z is a measure of how much the color of an object differs to an external observer as compared to the color observed by someone who is stationary with respect to the object. For moving objects, a value of z = 0 means that the two observers see the same color. As z increases, one observer sees the object as being redder than the other. According to the traditional Doppler shift, that means that one person sees the object moving away from them. A value of z = 1 means that one observer sees the object moving away from them at 60% the speed of light.

The variable z becomes tricky because not only does motion cause objects to appear to change color, so does the stretching of the universe, which is the cause of the color change of distant galaxies. In this instance, a value of z = 0 means that you are seeing an object in the present day, where a value of z = 1 means that you are seeing an object when the visible universe was half its current size.

In cosmology, it's the stretching of the universe that causes distant galaxies to seem redder, but we writers of popular science often say that it is because they are moving away from us. This is obviously technically wrong, but it's easier to imagine. And the distances are increasing, so . . .

Thus, what researchers did was to make a plot of the observed brightness of distant supernovae (which stands in for distance) versus the z variable (which stands in for the amount the universe has stretched or, more intuitively, the velocity of distant galaxies). Figure 5.2 shows a plot of the data taken by the Supernova Cosmology Project. This is the kind of data that were used to determine the expansion history of the universe.

You may recall that there were three possible outcomes in the experiment. The universe could be heavy, which means that the universe would slow very quickly. The universe could be light, which means that the universe would slow very slowly. And then

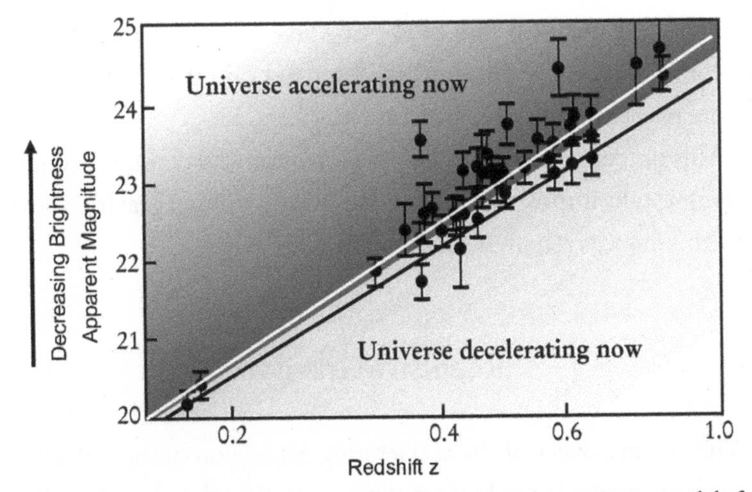

Figure 5.2 Data reported by the Supernova Cosmology Project. Redshift is a measure of time (and distance), where z = 0 is now and z = 1 is 7.7 billion years ago. Increasing apparent magnitude means dimmer. The white line is a fit to the data, while the black line is expectation if the universe only decelerated over its entire history. The data clearly favor a universe that is currently accelerating. (Figure courtesy SCP.)

there was the "just right" option, in which the universe slowed at a middling rate, leading to a stationary universe in the cosmos's far, far future.

So what did they see? One, two, or three? And the answer is . . . four. The expansion of the universe isn't slowing down—it's speeding up.

The expansion of the universe is speeding up in the present day, but that wasn't always so. The universe is about 14 billion years old. For the first 9 billion years of its existence, the universe was slowing down due to the attractive nature of gravity. This is what was expected. However, about 5 billion years ago—shortly before the birth of the Earth and our Sun—the situation changed and slowing transitioned to speeding up. Over the last 5 billion years, the expansion rate of the universe continues to increase. Metaphorically, the universe has stomped on the gas pedal, and there is a speeding ticket in its future.

This outcome was a shock to the astronomical community and seems totally impossible, given what we know about gravity. Some rethinking was required. How can this be?

Repulsive Gravity

With the discovery of the accelerating expansion of the universe, scientists began to explore what it meant. Was the idea of the Big Bang wrong? Did they just not understand how gravity worked? Was there a mistake in their measurements?

These and many other options have been considered, and some scientists still explore all viable explanations; however, the majority of cosmologists have selected one option as the most likely. This option makes no radical changes in our understanding of the laws of nature. Einstein's theory of general relativity and the Big Bang paradigm remain intact. To retain otherwise successful theories, cosmologists have postulated a new form of energy, called dark energy.

Dark energy, despite the similarities in the names, is nothing like dark matter. Dark matter is thought to be an unknown form of matter that interacts gravitationally in ways that are indistinguishable from the gravitational interactions of ordinary matter. In contrast, dark energy is simply a form of energy that permeates all of space and is essentially a repulsive form of gravity. (And, by repulsive, I mean it pushes things apart. I pass no judgment on what it looks like. That would be rude.)

The name "dark energy" was coined by American cosmologist Michael Turner at a conference in 1998. Michael and I have been on radio shows and discussion panels together, and I've personally heard him take credit for the term, as well as validating the claim with others who were present at the conference. So we know who to blame for the potential of dark matter/dark energy confusion.

What do we know about dark energy? From one point of view, we know very little. However, within the context of general relativity, we actually know a great deal.

Albert Einstein is, of course, generally conceded to be one of the most influential scientists of the twentieth century. Within the context of discussions of dark energy, it is his theory of general relativity that is most important. Simply speaking, general relativity is a theory of gravity.

Shortly after Einstein first presented his theory of gravity to the world in 1915, he was aware that the model contained within it what seemed to be a fatal flaw. Because of the attractive nature of gravity as it was understood at the time, his equations predicted that the universe would collapse under the impetus of gravitational attraction. Eventually, and over cosmic times, the universe shouldn't look like what astronomers of his time claimed.

At the time, Einstein was a believer in a static universe—one that existed forever, never changing on average. Because 1915 was before we understood a lot about stars, galaxies, and the universe as a whole, he wouldn't state his position in the way we would today. Sort of blending his position and modern knowledge, he believed that stars would be born and die in a never-ending cycle of birth, death, and rebirth—never changing and always looking the same. However, his theory predicted something quite different. So Einstein did what any scientist should do. When faced with a disagreement between his theory and the best data of the time, he modified his theory.

At its most basic, Einstein's theory of general relativity is quite simple. It's an equation that has on one side the distribution of energy and matter in the system being described—in this case, the universe. On the other side of the equation is the geometry of space and time. And, like most good equations, the two sides were joined by an equals sign. Stated simply, Einstein's equation says that matter and energy equals the curvature of spacetime.

In order to modify his theory so that it predicted a static universe favored by astronomers in 1917, Einstein added to the mass/energy side of the equation another term for which he used the Greek letter lambda (Λ). Λ is called the cosmological constant. A hasty person would also call Λ dark energy, although, as we will see, this is premature.

In the simplest form, Einstein's modified equation can be written as:

(regular mass and energy) + (cosmological constant) =
(spacetime geometry)

Regular mass and energy and the cosmological constant have an opposite sign, which means, while the gravitational force of regular matter and energy pull objects together, the cosmological constant pushes things apart. If the two are balanced properly, the attractive and repulsive forces cancel one another, which results in a static universe, just as Einstein imagined.

This modified equation was state of the art until 1929, when Edwin Hubble announced his observation that the universe was expanding. Einstein's preconception that the universe was static was wrong. And, in 1931, Einstein again did what any good scientist would do. In the face of data, he removed the cosmological constant from his equations. It is often said that Einstein said that the cosmological constant was his biggest blunder, although this statement is not found in his writings; rather, it is reported by individuals with whom Einstein spoke. However, whether the oft-quoted phrase is accurate or not, there is no question that Einstein considered the era in which he championed the cosmological constant to be an unfortunate one.

When Einstein died in 1955, he had reverted to his pre–cosmological constant equations. However, had he been alive in 1998 when the accelerating expansion of the universe was uncovered, there is no question that he would have resurrected the cosmological constant once again.

So what, exactly, is the cosmological constant? It is constant energy density, with a value equivalent to a mass of about 7×10^{-30} grams per cubic centimeter. This is very tiny. You can compare it to the average density of gas in interstellar space, which is 2×10^{-24} grams per cubic centimeter, or about 300,000 times denser than dark energy. Thus, inside galaxies, dark energy is entirely overwhelmed by the existence of ordinary matter.

However, galaxies—no matter how big they are—are just a minuscule part of the volume of the universe—tiny islands in a vast, intergalactic ocean. And the gas densities in intergalactic space can be very small indeed.

When one adds up all of the energy represented by the cosmological constant in the entire visible universe, it makes up 69% of the mass/energy budget in the universe. This is to be contrasted with the 5% occupied by ordinary matter and a 26% contribution of dark matter. At the present time, dark energy (represented by the cosmological constant) is the dominant form of energy in the universe.

But that wasn't always the case. Note that the cosmological constant is a constant *density*, not overall energy. It doesn't change as the size of the universe changes. This is in stark contrast to the behavior of both ordinary and dark matter, which are a fixed amount, which is to say, a certain amount of ordinary and dark matter was created when the universe began and that hasn't changed.

Since density is matter (or energy) divided by volume, as the universe has expanded, the density of ordinary and dark matter has gone down, while the density of dark energy has remained constant. In the last chapter, we learned a little about the cosmic microwave background radiation, which represents the conditions of the universe when it was 380,000 years old. At that moment, the radius of the visible universe was a little over a thousand times smaller than it is now, meaning the volume was about a billion times smaller. At that ancient time, dark energy was wholly negligible. Ordinary matter (including neutrinos and photons) was about 37% of the energy budget of the universe and dark matter was 63%. Atoms of ordinary matter made up only about 12% of the energy of the cosmos (and thus photons and neutrinos were about 25%). In this early universe, when dark energy was insignificant, the expansion of the universe is slowed by the gravitational attraction of ordinary and dark matter.

However, the universe expanded, with a corresponding increase in volume. Figure 5.3 shows how the density in matter and

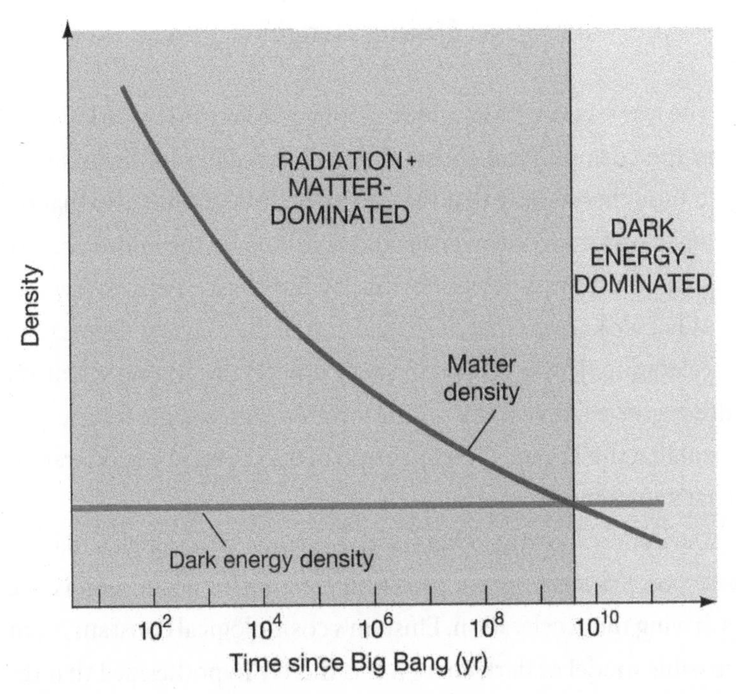

Figure 5.3 If dark energy is the cosmological constant, its density is constant, independent of the volume of the universe. However, as the volume of the universe increases, the density of matter decreases. About 5 billion years ago, dark energy became the dominant energy in the universe.

dark energy changed over time. At about 9 billion years after the cosmos began—about 5 billion years ago—the mix of matter and dark energy was about equal. And, since that singular moment, dark energy has begun to dominate. That, of course, means that the repulsive side of gravity won the battle and, as the universe continues to expand, dark energy is becoming ever-more dominant, causing the expansion to speed up. The cosmos is a runaway train, and there's no way to stop it.

Quintessence

In the last section, I was a little sloppy in referring to dark energy and the cosmological constant, and it wouldn't be surprising if you thought not only that the two were the same but also that the science was noncontroversial and accepted by the entire cosmological community. However, this is a little hasty. Let's review.

What we know is that the expansion of the universe seems to be accelerating. This is accepted nearly universally among scientists, although we'll revisit this a little later. So let's assume for the moment that this is true. The expansion of the universe is accelerating. That's an observational fact.

Dark energy, on the other hand, is a conjecture, albeit a reasonable one. Dark energy is a placeholder name for whatever it is that is driving the acceleration. Einstein's cosmological constant is one possible model of dark energy. It is merely hypothesized that the cosmological constant—Einstein's constant energy density with repulsive gravitational characteristics—is dark energy. That might not be true.

For one thing, the cosmological constant is . . . well . . . constant over all of space and all of time. However, it is possible that the energy field driving the accelerating expansion of the universe isn't constant at all. Perhaps it changes over time.

Cosmologists have a name for a form of dark energy that changes over time. They call it quintessence, from the Latin *quinta essential*, or fifth element. The name originates from Aristotle, where he postulated an element beyond the classical four of earth, air, water, and fire. Quintessence—his fifth element—was thought to be pure and associated with the heavens. Given the fact that dark energy seems to be a property of space, then the name quintessence seems at least poetically appropriate.

The cosmological constant is a fixed concept, but quintessence isn't. Any hypothetical form of dark energy that changes over time is a possible form of quintessence. So how do we determine if dark energy is constant or changing over time?

Well, the easiest—or at least most straightforward—is to repeat the measurements of the High-z Supernova Search Team or the Supernova Cosmology Project at larger distances. Astronomically speaking, large distances correspond to the distant past, so by looking at more distant supernovae, researchers will be able to get a better idea if dark energy has been changing over time.

The initial observation of dark energy used supernovae out to a distance of $z = 0.9$, which is a time when the visible universe was about 50% its current size and about 6.5 billion years old (about 7.5 billion years ago).

The most distant Type 1a supernova observed to date was at a distance of $z = 1.91$, which was about 10 billion years ago and when the visible universe was about a third its current size. Of course, this is a single supernova, from which it is difficult to generalize.

In the years since 1998, other experiments have looked to the heavens, searching for distant Type 1a supernovae. The Dark Energy Survey (DES) collaboration uses the Victor M. Blanco telescope, located in Cerro Tololo, Chile, to study approximately one-eighth of the entire sky. They have seen and catalogued hundreds of supernovae, with the majority being in the same distance range seen by the High-z Supernova Search Team and the Supernova Cosmology Project, with a few at somewhat larger distances. With the greatly increased number of supernovae observed and somewhat increased distances, the DES collaboration has concluded that the data are consistent with a constant dark energy density over the period of time for which they have data.

So does that mean the situation is resolved? Well, at some level, yes. The data agree with the cosmological constant, and there is no indication that dark energy changes over time. (Remember that dark energy is a constant energy density, not total energy. Total energy increases.) However, we only have respectable data from about 8 billion years ago to the present time. Observations of supernovae do not constrain the constancy of dark energy before that time or from this moment into the future.

It turns out that the tiny temperature variations we see in the cosmic microwave background is sensitive to the amount of dark energy in the universe, and measurements of the cosmic microwave background (CMB) also favor a constant density of dark energy over the lifetime of the universe.

However, let's imagine what would happen if dark energy is changing as a function of time. In order to be consistent with known measurements, the change would have to be small from when the universe began until now. However, it is entirely possible that, in the future, dark energy will either increase or decrease or even reverse its effect and become attractive instead of repulsive.

If dark energy remains the constant level it seems to have been for billions of years, the universe will continue to expand, with the rate accelerating over time. Distant galaxies will appear to move ever faster until the expansion of the universe effectively causes them to move away from the Earth at speeds faster than light. When that happens, those galaxies will disappear from view.

As time goes on, the accelerating expansion will cause even closer galaxies to slip away forever. Eventually, the entire universe visible to humanity will consist of the stars in the Local Group, a group of galaxies consisting of the Milky Way, Andromeda, and a bevy of nearby, smaller satellite galaxies. All other galaxies will be carried out of sight by the expanding universe.

If dark energy is increasing, then the effect will be more pronounced. The expansion of the universe may be sufficient to pull apart the Local Group. Even though Andromeda and the Milky Way will merge in a few billion years, the satellite galaxies might slip away. And, depending on how much dark energy increases, it's possible that the stars of the Milky Way will also disappear, as it is ripped apart by expansion far greater than we see today.

And it could get worse. The accelerating expansion of the universe might be enough to tear apart the solar system, the Earth, and even individual atoms. With enough of an increase in dark energy, in the distant future it could be that individual atoms float around, isolated from one another by cosmic distances. This bleak scenario is referred to as "The Big Rip," which is the Big Bang on steroids.

On the other hand, it could be that dark energy "turns off" in the future, leaving the universe to continue to coast, expanding ever more slowly. This scenario is roughly equivalent to the expectations of the cosmological community prior to the observation of the accelerating expansion of the universe.

Then there's an even more bizarre scenario—one in which dark energy becomes attractive rather than repulsive. Of course, the details depend a great deal on exactly how dark energy transforms itself; however, in this instance, perhaps the combined gravitational effect of ordinary matter, dark matter, and this new form of attractive energy could reverse the motion of the universe, pulling everything back together in a Big Crunch.

The fact that there is no direct evidence for the proposition that dark energy changes over time, that is to say quintessence, suggests that some of these possibilities are merely idle speculation. However, there a few known or theorized phenomena that keep the quintessence idea alive.

The first is wholly theoretical. Astronomers believe that spacetime is essentially flat. There are countless ways in which space can be curved and only one way in which it can be flat. On the face of it, the observed flatness of space is rather surprising. To explain this mystery (among others), American theoretical physicist Alan Guth proposed in 1980 the idea that, very early in the universe, the cosmos expanded at superluminal speeds. This super-fast expansion is called inflation and, so the thinking goes, in the breathtakingly short period of time of about 10^{-36} to 10^{-32} seconds after the Big Bang, the visible universe expanded from about the size of a hydrogen atom to a sphere about a light-year across.

Inflation has not been proven to be true, but it solves a number of cosmological mysteries. Relevant to the dark energy question, inflation is very much like dark energy. It was caused by a repulsive form of gravity that turned on briefly and then turned off. If inflation actually occurred and that repulsive gravity can turn on and off, then perhaps dark energy can as well.

While inflation is still a speculative idea, there is an observational astronomical conundrum that remains unexplained and could imply changing dark energy or something like dark energy.

Back in 1929, Edwin Hubble determined that the universe was expanding and that it was expanding at a specific rate, called the Hubble constant. The Hubble constant is a measure of how fast galaxies are moving away from us for each distance—at a certain distance, they move away at one speed; double the distance and you double the speed, and so on. Hubble and those that followed him measured the Hubble constant by looking at relatively nearby galaxies, determining their distances from Earth and their recessional velocity, and thereby worked out a relationship between distance and velocity. Because astronomers use relatively close galaxies, the Hubble constant is the expansion rate of the universe in the present day. While Hubble's initial determination was imprecise, a modern value for the Hubble constant is 74.03 kilometers per second of recessional velocity for every million parsecs of distance from Earth, with an uncertainty of 1.42 km/s/Mpc.

It turns out that there is another way to determine the Hubble constant, and this is to use the CMB, which, we recall, is a snapshot of the universe a scant 380,000 years after the universe began, and then use our knowledge of cosmology to predict the Hubble constant in the present day. When astronomers do this calculation, they come up with a value of the Hubble constant of 67.4 km/s/Mpc, with an uncertainty of 0.9 km/s/Mpc. The two numbers do not agree, not even when the uncertainties are taken into account. This 10% discrepancy could well constitute a real problem.

Basically, the technique is like taking a baby's height and predicting their height when they are full grown, and then comparing that prediction to their actual adult height. If you understand how people grow, you could do this. And astronomers think they understand how the universe should expand, so this discrepancy is a real mystery.

Now the obvious resolution to what is called the Hubble tension is that one of the two measurements is either wrong or has underestimated their uncertainties. However, both groups of researchers are among the best in the business, and many others have looked over their analysis with a fine-toothed comb. No obvious errors were committed.

So what could the answer be? What could explain a measurement taken early in the history of the universe predicting a smaller rate of expansion than we observe today? Well, one possible explanation is that at some time early in the lifetime of the cosmos—after the universe cooled enough to let light travel unimpeded at 380,000 years after the Big Bang, but not too long after that, perhaps there was some form of "dark energy" (which could be the same as or something completely different from the dark energy driving the accelerating expansion of the universe today). This short period of acceleration would give the universe a "kick," so to speak, and the universe today would be expanding faster than is predicted using measurements of the CMB.

Now, to be honest, this idea that there was some form of energy akin to dark energy that turned on, then off again, about a million years or so after the Big Bang is an unlikely one. But the bottom line is that we don't know.

Thus, the possibility of inflation turning on and off and maybe supplying the "kick" that resolves the Hubble tension at least suggests that the idea of quintessence might be right, albeit with a change on much longer timescales. Neither inflation, nor the kick resolving the Hubble tension, should be considered as real evidence for quintessence. But they at least point to the possibility that quintessence is real.

The Worst Prediction in Science

While the evidence for quintessence is sketchy at best, there are some other unanswered mysteries of dark energy. But there is one that is really a doozy.

If dark energy is real and the cosmological constant is the correct explanation, then dark energy has a value equivalent to a mass of about 7×10^{-30} grams per cubic centimeter, give or take a few. Furthermore, it appears that dark energy is a property of space. As space expands, dark energy appears to be constant, and it is this constancy that makes scientists believe that space and dark energy are so inextricably intertwined.

In Chapter 2, we learned about the standard model and how it all worked. In essence, an extended form of quantum mechanics governs the behavior of the microcosm.

One of the best-known principles of classical mechanics is that energy is conserved, meaning that total energy never changes. However, quantum mechanics breaks that rule. In the quantum realm, the Heisenberg uncertainty principle applies, which basically states that energy doesn't have to be conserved, as long as the violation of energy conservation doesn't take that long. Basically, energy can differ from the classical expectation. If the difference is only a little, that can last a "long" time (still less than a second), but if the difference is large, the discrepancy must disappear blindingly fast.

The consequence of Heisenberg's principle and the quantum fields is that pairs of matter and antimatter quarks and leptons can briefly appear and disappear, although you can't see it because the process happens just too quickly.

It turns out that it is possible to use advanced quantum theory to predict the energy inherent in this process. And, since this process exists everywhere in space, it is often thought that this quantum energy is dark energy.

That would be an extraordinary intellectual coup, if it were true; the energy we see driving the expansion of the cosmos is the same energy that we have observed in quantum mechanics. It would provide a very clear path forward toward linking quantum mechanics and gravity—the standard model and general relativity. In short, it would be a powerful step toward creating a theory of quantum gravity and eventually a theory of everything. Except it doesn't work.

If we calculate the energy predicted to be contained in the quantum fields, we find that the prediction is a mind-boggingly huge number of 10^{120} times bigger than dark energy. That's a big one, with 120 zeros after it. This has been called the most staggeringly bad prediction in all of physics.

This outrageous disagreement of 120 orders of magnitude difference between quantum predictions and gravitational measurements means that we very clearly don't understand something. *Something* has to cancel that huge number and reduce the energy of space to that we measure when we study the cosmos. We just don't know what. However, when this question is resolved, it will be a huge advance in our quest of a theory of everything.

Other Possibilities

The majority of the cosmological community is convinced that the expansion of the universe is accelerating, and that Einstein's theory of general relativity applies. That means that dark energy in the form of the cosmological constant is the best explanation so far. However, not everyone accepts these conclusions. These contrarians think that there are better explanations.

One faction in the camp of dark energy skeptics is not convinced that the data support a conclusion that the expansion of the universe is accelerating. They raise several points that they think have been overlooked by proponents of dark energy.

One claim has to do with a technical point regarding how the supernova data were analyzed. Basically, the data are comprised of a few dozen supernovae, with each supernova being described by a distance and a velocity (or apparent magnitude and red shift). Scientists then fit that data and extract information from the fit parameters.

However, different scientists can fit the data points with different mathematical functions and get different results. For instance, in Figure 5.4, we see some made-up data, each with values for distance and velocity. We also see two fits for the data—one a straight line and one curved. Depending on which curve is used to represent the data, one will make different conclusions.

For most researchers analyzing supernovae, they will use general relativity theory to guide how they extract meaning from their data. However, there are others who claim that assuming

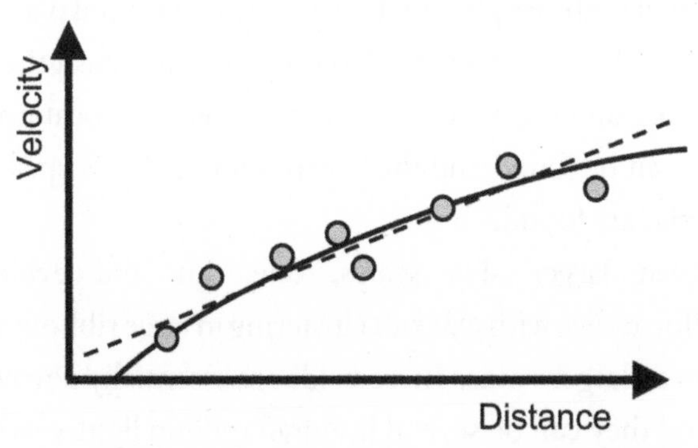

Figure 5.4 An example of data being fit to a straight line (dashed) and curved one (solid). Curved lines require more parameters to describe, which usually means that a curved line will better represent data than a straight line.

general relativity to be true will bias one's conclusions and thus they make different choices; therefore, the two groups each claim that the other group is making a fatal mistake. While the majority of the cosmology community uses the equations of general relativity to perform their analysis, there remain a small number of noisy contrarians, and the conversation continues.

Most of the arguments over the topic of whether the universe is accelerating or not hinge on the assumption of a uniform universe. This is called a Friedmann universe. Alexander Friedmann was a Russian physicist and, in 1922, he used Einstein's theory of general relativity to model the universe as a whole. However, in his formulation, he made the simplifying assumptions that on the very largest scales the universe was isotropic and homogeneous—which is just a fancy way to say that the universe is smooth, with no density variations.

On the face of it, the claim that the universe is isotropic and homogeneous is a silly one. After all, we can look to the night sky and see that there are places where there are stars and places where there aren't. Using powerful telescopes, we can repeat the process for galaxies, and the result is the same. There are locations where galaxies can be found and then there's intergalactic space, where no galaxies are found.

At even larger size scales, one can find evidence of nonuniformities, with galaxies clustering in large ribbons that surround even larger voids. These voids are essentially devoid of galaxies, and they can be several hundred million light-years across. Surveys of the location of galaxies out to a few billion light-years show that the universe has a structure similar to a pile of soap bubbles, with galaxies found mostly in the area of soapy skin and not in the voids.

However, at even larger size scales, the universe seems to be uniform. In the same way that the foam on a pint of Guinness (or root beer if you prefer) appears to be bubbly if you look at it closely and smooth when you hold it arm's length, so, too, it is with the universe. Figure 5.5 shows the structure of the universe at size scales that both emphasize the foamy nature of the cosmos at smaller scales and the uniform nature at larger ones.

However, by studying the CMB, which is the afterglow of the Big Bang that persists in the modern day, astronomers have determined that the entire universe is much, much, bigger than the universe we can see. The visible universe has a diameter of about 92 billion light-years across. (Naively, you'd think that we can only see as far as the lifetime of the universe times the speed of light. Given that the universe is about 14 billion years old, that would suggest that the visible universe should have a diameter of only 28

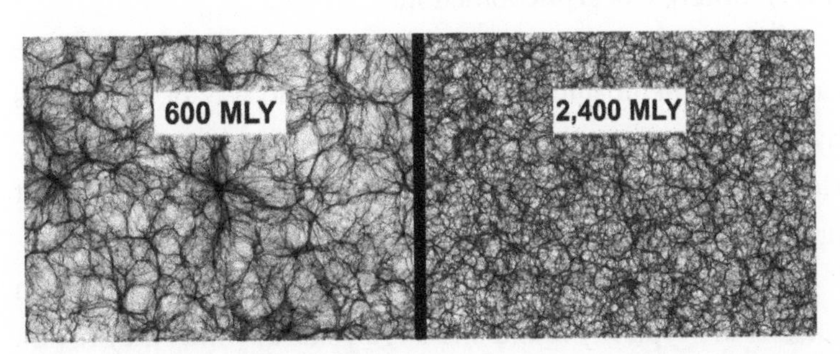

Figure 5.5 The large-scale structure of the cosmos. The left side is magnified more than the right. The length of the white bars on each figure corresponds to 600 and 2,400 million light-years. These results are simulated. (Figure courtesy of the Millennium Simulation Project.)

billion light-years across. The much larger size is because of the expansion of the universe.)

While the diameter of the visible universe is about 92 billion light-years, the diameter of the entire universe is much larger. At a minimum, it is at least 500 times larger than that—and it could be much larger. Taking the minimum, the entire universe is at least 46 trillion light-years in diameter, with a volume that is at least 125 million times larger than the universe we can see.

Given that the universe is so much larger than we can see, it is possible that there are structures in the cosmos that are even bigger than our visible universe. If that's true—and I need to emphasize that we have no evidence that it is true—maybe our visible universe is in a region of lower density than the average density of the universe. If that's true, then the accelerating expansion of the universe is simply because the part of the universe that we can't see is denser than our familiar surroundings and that is the reason that the expansion is speeding up. Essentially, we live inside a bubble that is expanding. If we could see the entire universe, we'd see that this is entirely a local phenomenon.

Talking about a cosmos that is bigger than the visible universe can beggar the imagination, and we have no way of testing this hypothesis, so this idea will have to remain speculative conjecture. However, the structure of the universe we *can* see is also brought up by those who are suspicious of dark energy.

As we've seen, the visible universe consists of filaments in which galaxies are typically found and large voids in which they're not. Since Type Ia supernovae are stars and stars are found in galaxies, that means that the data that astronomers record are biased. This means that the measurements occur in denser parts of the universe than average. (For instance, the volume encompassed by intergalactic voids is about 80% the volume of the universe.) Accordingly, some researchers claim that the expansion of the universe varies whether one is looking at space in a void or in a galaxy cluster, and this is not taken into account when researchers analyze the data.

And so the debate unfolds. However, it is important to remember that among mainstream cosmologists, the discussion is largely settled; dark energy is real and constant or nearly so.

However, what sorts of data will help strengthen that position? Well, there are some experiments that will take data in the next decade or so that might help. The Vera C. Rubin Observatory (previously called the Large Synoptic Survey Telescope, or LSST) has many goals, but one will be to survey deep into the universe to better constrain dark energy models. It is located in Chile and is scheduled to begin operations in October 2023, although schedules often slip in such projects.

In contrast, the Dark Energy Survey Instrument (DESI), located in the southwest United States, saw first light in 2019. It is designed to make precise three-dimensional maps of the universe and will help validate or refute both dark energy and general relativity. As of this writing, no results have been released, so we'll just have to wait and see what secrets of the universe it will reveal to us.

Then there is the Euclid satellite, being built by the European Space Agency. This facility is scheduled for a launch in 2023, although, as with all such schedules, *caveat emptor*. It will also try to refine our understanding of the expansion history of the universe.

In short, we are still trying to understand the nature of dark energy. While the general consensus is that it exists, there remain some diehard holdouts who believe otherwise. If their way of thinking gains traction, we'll have to significantly rethink our understanding of the cosmos. Then there's the very big elephant in the room—the one where a prediction and measurement of the amount of dark energy in the universe disagree by a whopping 120 orders of magnitude, which is a staggering discrepancy, suggesting that we are missing something huge.

And, of course, there is the unanswered question of whether dark energy is constant or not. Depending on the answer, the future of the universe could be very, very different. We still have a lot to learn.

In this chapter and the last, we've learned of the many unanswered questions of the dark universe. Dark matter and energy make up a full 95% of the energy budget of the universe, suggesting that maybe we've been studying the wrong things. Rather than getting a good handle on the behavior of ordinary matter and energy, we need to better understand the most prevalent components of the cosmos. Big picture wise, we've only begun to scratch the surface of the makeup of the universe. However, we're not done in our attempt to catalogue our list of things that we don't yet understand. In the next chapter, we'll talk about how even our understanding of ordinary matter and energy is quite lacking and how we can't explain why we're even here at all.

CHAPTER 6

MISSING ANTIMATTER

In the last two chapters, I talked about big mysteries—ones in which fully 95% of the matter and energy content of the universe remains elusive and not understood. However, in Chapter 2, I told you about the great successes of the standard model and how we have a good understanding of the behavior and nature of ordinary matter. There's just one problem. There is an enormous mystery governing the behavior of familiar matter and energy. And it's not the piddling factor that arises from dark matter and energy being so much more prevalent than ordinary matter—it's something where the ratio of observed to missing matter is more than a billion to one.

The most famous equation of physics, perhaps in all of science, is Einstein's $E = mc^2$. Basically, it says that matter can be converted into energy and vice versa. While that's essentially true, it's not the entire story.

In Chapter 2, I introduced the concept of antimatter, which is a sibling substance of matter. However, like many sibling relationships, the rapport between matter and antimatter is a fiery one. Combine equal amounts and types of matter and antimatter, and the two will annihilate and convert into energy.

Einstein's Unfinished Dream. Don Lincoln, Oxford University Press. © Oxford University Press 2023.
DOI: 10.1093/oso/9780197638033.003.0006

And the opposite process is also true. Energy can be converted into matter, but always with an equal amount of identical, but opposite, antimatter. For example, an energetic photon can, under the right conditions, convert into an electron and antimatter electron (i.e., a positron). Photons can also convert into matter/antimatter pairs of quarks. And other forms of energy can experience the same conversion. Figure 6.1 shows examples of the kinds of processes that can occur.

The key point is that when energy converts into matter and antimatter, the amount of matter and antimatter are identical. If you're a math-minded person, if you think of matter as being +1, antimatter being −1, and energy being 0, it all makes sense. After all, +1 + −1 = 0. It's very straightforward, and this balance in the creation of matter and antimatter has been tested to outrageous precision—they're always created equally.

In and of itself, this does not pose any sort of problem. However, a mystery arises when one combines this matter/antimatter symmetry with the idea of the Big Bang. As you recall, the Big Bang hypothesizes that the mass of the visible universe was originally compressed to a tiny point of essentially zero size. That microdot began to expand, and the conditions of the early maelstrom were

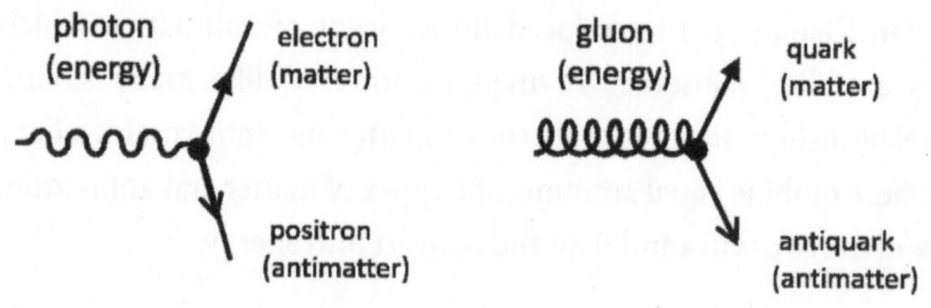

Figure 6.1 When energy like a photon (left) or gluon (right) converts into matter, it makes an equal amount of antimatter.

unimaginably hot. The entire visible universe (and beyond!) was filled with an enormous amount of energy.

We've already seen what can happen when energy is present. It makes matter and antimatter in equal quantities. And yet when we look out at our universe today, we only see matter. Basically, antimatter is not to be seen. So that leads us to a very vexing question.

Where is the antimatter?

Actually, the real conundrum is a little more complex, although that simple question contains the crux of the matter. However, to be a bit more accurate, what should the universe look like today if the Big Bang and the matter/antimatter symmetry were the entire story?

Well, in the early universe, energy would have made equal amounts of matter and antimatter. And, because the process goes both ways, the matter and antimatter would encounter each other and convert back into energy. In a static universe, what we would expect to see would be a hot soup of matter, antimatter, and energy, all sloshing back and forth into one another. That's for a static universe.

However, we know the universe isn't static. Because of the Big Bang, the universe is expanding. That has two super important consequences. The first is that the conditions are constantly changing. If, say, a photon created an electron and positron in an expanding universe, then when it came time for the electron and positron to convert back into a photon, the universe would be slightly bigger. That means the two particles would be farther apart, and this would make it marginally more difficult for the two particles to find one another and annihilate. So the expansion of the cosmos slightly favors energy converting into matter and antimatter, at least when one considers the simple case of one energy particle and two matter/antimatter ones.

Of course, the universe was full of energy and, consequently, full of matter and antimatter. It's not required that the electron and positron that were created in the example I just described annihilate with each other. Indeed, *any* electron and positron could annihilate. So even in the expanding universe, the annihilation would continue. However, since the expanding universe would decrease the density of matter and antimatter, the annihilation rate would slow down, as it would be harder and harder for a matter and antimatter particle to encounter one another and annihilate. That would suggest that the cosmos of today should be a bunch of matter and antimatter particles, separated by large enough distances that they rarely encounter one another to annihilate.

However, there's another ingredient in the mix, and that is the effect of the expansion of the universe on energy. It is well known that photons, being a form of light, have a wavelength, but this is also true of all forms of subatomic energy. And the expansion of the universe plays havoc with those wavelengths. As time goes on, the stretching of space also stretches the wavelength of energy particles. Since short wavelengths imply high energy and long wavelengths mean lower energy, this means that the amount of energy held in these subatomic energy particles decreases over time. (And, yes, that does mean that energy is not conserved in the familiar way in a universe that is changing its size. This is a long-known consequence of general relativity, and a thorough exploration of the consequences of this would divert from the narrative.)

The changing of subatomic energy due to the expansion of the universe has significant repercussions. Let's explore why with a concrete example. An electron has a specific mass, specifically

0.511 million electron volts (MeV), when converted to energy units. If you prefer mass units, the electron's mass is 9.1×10^{-31} kilograms, and we can convert between the two using $E = mc^2$.

The positron has identical mass. Thus, to make an electron/positron pair, you need 2×0.511 MeV of energy, or 1.022 MeV. So what does this mean for the early universe? Consider the following scenario.

Suppose you have a photon with 1.025 MeV of energy. This is enough energy to make an electron and positron. These two particles exist for a short time and then recombine, recreating a photon with 1.025 MeV of energy. Now, while the photon exists, the universe expands, reducing the photon's energy to 1.021 MeV. This energy is too low to make an electron/positron pair, and so the photon can no longer create matter and antimatter. Once the energy of the photon is below the threshold required to make an electron and positron, the creation of matter and antimatter stops, and the photon continues to experience the expansion of the universe. This means that the photon would slowly lose energy year after year.

OK, so with all that prologue in place, what would we expect to happen if the experimentally observed symmetry in matter and antimatter were combined with the straightforward consequences of the Big Bang? We would begin with a cosmos full of energy. That energy would create matter and antimatter, which would then annihilate back into energy. The density of matter and antimatter would reduce, although the fact that it was so thoroughly mixed means that the annihilation would continue. However, the energy of individual subatomic energy particles would reduce until eventually they could no longer convert into matter and antimatter.

Thus, we'd be left with a cosmos full of photons of energy that were slowly losing energy over time, interspersed with a rarefied gas of electrons and positrons, as they are the lightest and stable known subatomic particles. And, over the eons, electrons and positrons would continue to have chance encounters and annihilate. The universe today would be a bath of low-energy photons, with a rare leavening of electrons and positrons, and the electrons and positrons would exist in equal numbers. And that's all well and good, except for an inconvenient fact, which is that this doesn't describe the universe in which we live. And that, as they say, is a mystery.

The Actual Universe

We've explored what our universe should theoretically look like. Now let's take a look at our actual universe, viewed through the same eye.

As we stare out into the night sky, looking at the familiar vista of stars and galaxies, planets and comets, the first thing we note is that there's something actually there. The universe isn't a bath of energy, infused with a wispy and diffuse gas of matter and antimatter. We see clumpy matter—and only matter—made up of atoms. Using both optical and radio telescopes, we can figure out the amount of matter in the visible universe.

How about the antimatter and energy? Well, the antimatter is the easiest. Basically, we see none. Now we have to be careful. When I say none, I'm talking about antimatter moving at the speeds that are customary in astronomical studies—say a few tens or hundreds of kilometers per second, like the 30 km/s velocity of the Earth orbiting the sun, the 200 km/s velocity of the Earth orbiting the Milky Way, or the 600 km/s velocity of the Milky Way speeding through the universe.

However, astronomers have found instances of antimatter traveling at near relativistic speeds. These are the result of high-energy processes, like magnetic fields near pulsars, colliding neutron stars, or high-energy jets of matter being spewed from the poles of black holes. Antimatter in this form wasn't created in the Big Bang and isn't what we're talking about. Searches for antimatter originating from when the universe began basically come up empty. There appears to be no primordial antimatter in the universe, although we'll revisit this topic in an upcoming section.

What about a census of primordial energy in the universe? Where can one find that? We talked in earlier chapters about the cosmic microwave background, or CMB. The CMB is the vestigial signature of the fireball of the Big Bang and provides a compelling estimate of the energy content of the universe 380,000 years after the Big Bang. This, in turn, is a window into the energy content to even earlier times.

Shortly after the universe began, it was filled with energy and the swirling soup of matter and antimatter described in the previous section. Every time a matter/antimatter pair of particles was created, the number of energy particles decreased by one. Later, when the matter/antimatter particles annihilated, the number of energy particles increased. Eventually, the universe expanded and cooled enough that the energy particles had too little energy to make more matter and antimatter. We covered this in the last section.

Thus, the number of energy particles at the moment which the universe cooled enough to no longer create matter/antimatter pairs is a measure of the number of matter/antimatter pairs that existed at the time. Because quarks weigh more than electrons, the final energy particles to participate in the whole frenzied process were photons. This is because gluons can't make the light electron and positron. And, from that point onward, those photons continued to be absorbed and emitted by any matter remaining in the universe.

At first blush, you'd think that the number of photons would change in the plasma that existed at the time, but the temperature of the universe was unchanging, or rather changing very slowly; thus, any matter that existed at the time would, on average, emit as many photons as were absorbed. This process continued until about 380,000 years after the cosmos began, at which point it became cool enough for hydrogen atoms to form. At that point, the photons could then travel unimpeded across the universe.

These photons had a range of wavelengths, with the average being about one micrometer, which is infrared light. Over the intervening nearly 14 billion years between that moment and now, the universe has expanded, stretching those photons to a wavelength of about a millimeter, or what scientists call microwaves. And astronomers are able to use radio telescopes to detect these microwaves. The discovery of this microwave background occurred in 1964, and it is considered to be a smoking gun in support of the theory of the Big Bang.

However, what is most relevant for our purposes is that we can determine the number of photons in the CMB, which we can then trace back to the number of matter and antimatter particles in the universe when the last matter/antimatter annihilation was going on. We can then compare that number of photons to the number of protons in the visible universe, and we find that the photons outnumber protons by a factor of about 2 billion to one. (One must take that number with a degree of uncertainty. Different scientists, using different ways to take the census of photons and protons, making different approximations and rounding differently, will come up with estimates ranging from about 1 billion to 10 billion. Two billion is my attempt, but any number in the range of 1 billion to 10 billion is a reasonable estimate. For the rest of the text, I'll use the 2 billion number.)

So, using the 2 billion number, what the data are telling us is that shortly after the Big Bang, something occurred to make matter slightly more common than antimatter. For every 2,000,000,000 antimatter particles, there were 2,000,000,001 matter particles. The 2,000,000,000 matter and antimatter particles annihilated, with that single matter particle going on to make up the universe we see now. And, to be clear, I'm talking ratios here. There were many more particles participating than 2 billion.

OK, so now we're getting somewhere. Data are showing us that, early in the universe, matter was slightly favored over antimatter. And *that* is a mystery to unravel. Be warned that the next couple of sections are a bit technical.

Mirror, Mirror

In the 1950s, one of the pressing questions of particle physics was called the "tau-theta puzzle." This conundrum seems inconsequential and technical at first, but answering it had profound consequences for our understanding of the forces of nature and provided a first hint of an asymmetry between matter and antimatter. We're still working out the implications of this measurement today.

Conservation laws govern all subatomic interactions—actually all interactions, including those at classical sizes, for that matter. Energy and momentum are two conservation laws, which is just a fancy way to say that a quantity doesn't change. If you cause two billiard balls to collide and determine their combined energy before the collision, the sum of their energies after the collision will be the same number. This is also true for momentum. It's also true for angular momentum, which is a complicated mix of the mass, shape, and speed of spin of an object.

There are less familiar conservation laws that apply. Of the less familiar ones, conservation of charge is perhaps the easiest to visualize. Charge conservation means that the total charge of an isolated system never changes. If a photon (with electric charge equal to zero) converts into two particles, the sum of their charges must also be zero. For example, if one particle has a charge of + 1 unit, the other must have –1 unit.

Parity is a totally unfamiliar property that applies in the quantum realm, and it has to do with the wave function that governs the system. For a simply two-dimensional function, drawn in the x-y plane, a function is said to have positive parity if, when you swap the +x, direction with the –x direction, the function looks unchanged. That's a fancy way to say that you can flip the function left to right and not notice. If you flip the function and it's the opposite, we say this function has negative parity. And, of course, not all functions have positive or negative parity. For some, if you flip left and right, they look very different. Figure 6.2 shows examples of functions with positive, negative, and no parity.

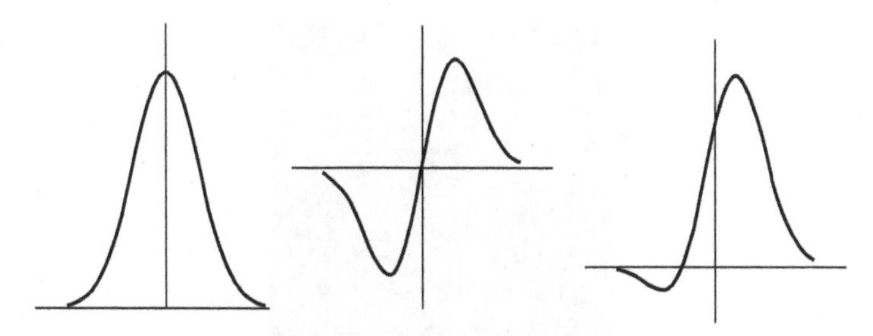

Figure 6.2 Functions with positive parity (left), negative parity (center), and neither (right). In all cases, the horizontal and vertical lines are the x and y axis for that specific plot.

Each subatomic particle has a specific parity, which is a compact commentary on its quantum mechanical wave function. One particular subatomic particle is called the pion, short for pi meson. It is the lightest of the particles that contain quarks. Pions happen to have a negative parity (e.g., $P = -1$). You'll have to trust me on that, but it's true.

OK, so what does this have to do with the tau-theta conundrum? Well, the tau (τ) and theta (θ) particles were two denizens of the dizzying menagerie of particles that were discovered in the 1950s. They were example of what were called "V" or "strange" particles. We now know that this means that they contain at least one strange quark, although that was not known at the time. They were classified as V particles because when they were photographed in detectors of the era, they left a V-like pair of tracks in the images. At the time, a strange particle is simply one that was easily created, but decayed slowly. Figure 6.3 shows examples of V particles.

The tau and theta particles seemed to be identical in essentially every way. They had the same electric charge, mass, spin, and lifetime. However, they differed in one important way. The tau particle decayed into three pions, while the theta decayed into two.

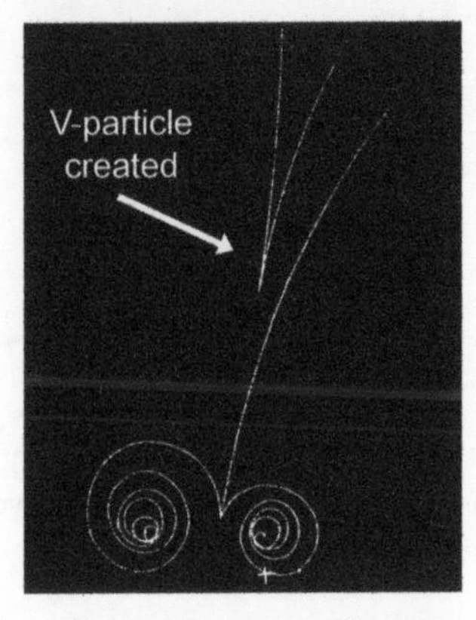

Figure 6.3 A particle collision that created a "V" particle, among other things. (Figure courtesy Lawrence Berkeley Laboratory.)

Now this small difference might not seem to be such a big deal to you, but they were incredibly puzzling to the scientists of the era. The problem had to do with parity. It was not possible to determine the particle's parity directly. However, the parity of their decay products (pions) was known, and, through conservation laws, the parity of the theta and tau could be inferred.

As I mentioned previously, the pion has a negative parity. When you combine multiple particles, the parity of the combination of the ensemble of the two or three particles can be determined by multiplying them together. Thus, the tau particle, with its three-pion decay mode has a negative parity, for example, $(-1) \times (-1) \times (-1) = -1$. In contrast, the theta particle has a positive parity, for example, $(-1) \times (-1) = +1$.

So this is very peculiar, and scientists were driven to two conclusions. The first is that the two particles were literally two different particles that were the same in every way except that they had different parities. The second was that the parity of the particle was not conserved. After all, if the two final states had either positive or negative parity, then whatever the initial particle's parity, half the time it got changed in the decay. Thus, there was a real possibility that parity was not preserved.

Now in the early 1950s it was taken as an obvious fact that parity was conserved. However, in 1956 physicists Tsung-dao Lee and Chen Nin Yang did a survey of the literature and found that this was still an open question in weak force interactions. They found extensive evidence that parity was conserved in both strong force and electromagnetic decays, but nobody had tested the principle in weak force decays. This led to their famous paper "Question of Parity Conservation in Weak Interactions." In it, they substantiated their claim that the question of parity conservation in weak forces had not been examined, and they suggested a few possible experiments that could answer the question.

One of the possible experiments mentioned in the paper had been suggested by Chien-Shiung Wu, the first woman to be granted tenure in the physics department at Columbia University. She and Lee had discussed the question in the spring of 1956. Lee was a theoretical physicist and Wu was an experimentalist. He proposed a few ideas, but then she suggested studying the decays of cobalt-60.

Wu didn't only suggest the experiment; she performed it in collaboration with a number of technical experts. Here's how it works. Cobalt-60 is a radioactive isotope that decays with a half-life of about five and a half years. It decays into nickel-60. Nickel has one more proton than cobalt, so what happens is that a neutron in the cobalt nucleus decays into a proton, electron, and electron antineutrino.

Cobalt-60 has a nuclear spin of 5. The nickel-60 into which it decays has a nuclear spin of 4. (That nickel-60 then emits some gamma rays, which further modifies the nickel nucleus's spin, which is why stable nickel-60 has a spin of 0. However, for purposes of the decay here, it is the spin of 4 that is important.)

In the decay, the spin decreases by one and spin (which is angular momentum) must be conserved. The electron and antineutrino both have a spin of ½, which means that when the two particles leave the interaction, they both must have a spin in the same direction (because (+½) + (+½) = 1). For the sake of simplicity, let's say that the spin of the cobalt-60 atoms is pointing in the upward direction. That means that the spin of the decay particles must also be up.

Two more things are important to know. The first is that for particles like electrons and neutrinos, the spin is either in the direction or opposite the direction of motion of the particle. The second is that in order to conserve momentum, which is a measure of the motion, if the parent is stationary, then the two decay products must head off in exactly opposite directions.

Putting this all together, here's what should happen. If you orient the spin of the cobalt nuclei so they point upward, then after the decay, one decay particle should shoot upward and the other downward. And, if parity is conserved, the electron should shoot upward and downward with equal probability. If anything else is observed, then parity might not be conserved.

Aligning the nuclear spins of a sample of cobalt required that it be cooled to incredibly low temperatures, specifically 0.01 kelvin, or −459.67° F. It also needed to be immersed in a strong magnetic field. The magnetic field aligns the spins of the cobalt nuclei, and the low temperature keeps the nuclei from moving enough to misalign their spins.

Wu didn't have the requisite experience in low temperature technology, so she enlisted collaborators from the National Bureau of Standards (now the National Institute of Standards and Technology) to help.

With their equipment in place, they performed the experiment. What they found was that if the spin of the cobalt-60 nuclei was pointed upward, then electrons were emitted downward, as shown in Figure 6.4. They couldn't detect the antineutrinos, but this meant that the neutrinos were emitted upward. The health of parity conservation in weak interactions seemed to be in peril.

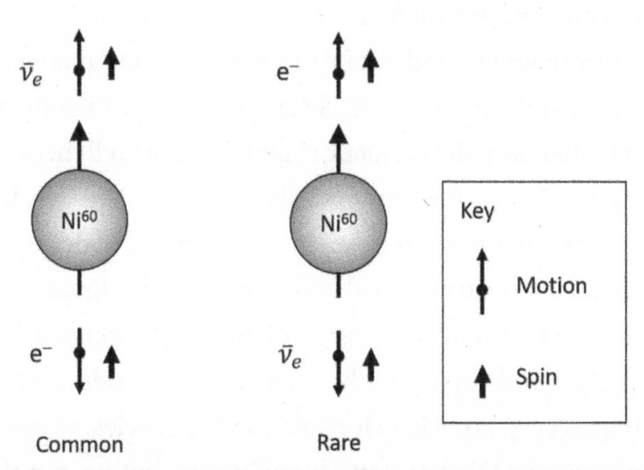

Figure 6.4 In the Wu experiment, cobalt-60 nuclei had their spin of +5 aligned in the upward direction. After the decay, the nickel-60 nuclei were oriented in the same direction, but with a spin of +4. Accordingly, the sum of the spins of the electron and antineutrino had to equal +1. The fact that electrons were only emitted downward showed that parity conservation was violated in weak nuclear decays.

Wu and collaborators were not the only people to test parity conservation in weak interactions. Wu and Lee were professors at New York City's Columbia University, which was (and still is) one of the premier physics departments in the country. However, there was another brash young faculty member there, by the name of Leon Lederman. He heard of Wu's work and reconfigured an existing experiment to perform a similar test using the weak force decay of pions and muons, and he came to the same conclusion as Wu did. Combining Wu and Lederman's observation clinched the conclusion. Parity conservation in weak force interactions was dead.

While the death of parity conservation sent shockwaves through the scientific community, for our purposes it had a more interesting consequence. It gave scientists a way to distinguish between matter and antimatter.

In electromagnetic and strong force interactions, matter and antimatter are indistinguishable. Swap all matter and antimatter in any particular particle interaction, and you can't tell them apart.

However, Wu's experiment showed that in weak force interactions, that wasn't true. In her cobalt-60 decays, all of the antimatter neutrinos had their spin in the direction of the antineutrino's motion. Scientists call this a right-handed spin.

Invoking handedness might seem to be an odd concept for spinning objects, especially for subatomic particles, where spin is a fuzzy concept. Objects don't literally spin, but we can think of them as spinning, as long as we don't pursue that line of thought too deeply. How does handedness come into play?

A spinning object, like a ball, has a motion associated with it. If you look along the axis of rotation, the object will either seem to rotate clockwise or counterclockwise. Now look at your hands.

Point your thumbs at your eyes and you see that your fingers naturally curl—the fingers of the right-hand curl anticlockwise, while the fingers of the left hand curl clockwise.

For a moving subatomic particle, you point your thumb in the direction the particle is moving. If it spins in the direction of the fingers that your right hand curls, we call that a right-handed particle. If it spins the other way, we call it a left-handed particle.

With that nomenclature in place, what Wu found was that antineutrinos are right-handed particles. Subsequent research has demonstrated that neutrinos are all left-handed. At least in weak force interactions, we can tell between matter and antimatter by determining the particle's spin direction.

Now I should probably take a moment and clarify a few things and focus the conversation on the key points. Many of these were points about which I was very confused when I first heard about this topic. If you're not much of a caveat person, you can safely skip to the next section.

First, the names of the theta (θ^+) and the tau (τ^+) were retired and reused. What once was the θ^+ and τ^+ are now known to be a single particle, called the K^+ meson. No physicist that was born in the last half a century or more uses the old nomenclature. Nowadays, the symbol θ is used to denote a meson consisting of a top quark and an anti–top quark. Given that the top quark decays so rapidly, it is expected that they will decay before a theta meson can be formed. Accordingly, there is no expectation that a theta meson will ever be discovered. Also, some scientists use the symbol θ to represent a rare form of matter called a pentaquark. Pentaquarks are subatomic particles that contain four quarks and one antiquark. (In contrast to the proton, which is a baryon with three quarks.) Pentaquark research is in its infancy. Pentaquarks were discovered at the CERN laboratory in 2015, and a few have been observed since then. However, early reports of the observation of the θ pentaquark have been ruled out by subsequent research. As of this writing, the θ pentaquark remains a hypothetical particle.

The symbol τ has been repurposed to represent the heaviest charged lepton—essentially a rotund cousin of the electron. It was discovered in 1975, and it is an established and well accepted member of the standard model.

All of this makes it tricky if you want to dig into historical literature to learn more about this interesting moment in physics history. Basically, only research papers published before 1957 or ones that are detailing the history of physics use the old terminology for θ^+ and τ^+.

Of more import is the implication of the Wu experiment. What this experiment and the follow-on ones implied is that the weak force only interacts with left-handed leptons and right-handed antileptons. Note that this *doesn't* mean that leptons are left-handed and antileptons are right-handed. In interactions involving the electromagnetic or strong force interactions, matter and antimatter can be either kind of handedness.

In addition, for particles with mass, handedness depends on your reference frame. Take, for example, a right-handed particle moving toward the right at some modest speed, say 10 meters per second. This particle has both its motion and spin pointing to its right. Now ask yourself what a person moving at 20 meters per second to the right will see. Since they are moving faster than the particle, they'll see it moving to the left, but the spin direction won't change. Accordingly, this second observer would call this a left-handed particle. Both observers are correct, which means one has to be careful about this whole business.

An important difference arises when talking about neutrinos. Neutrinos are nearly massless, which means that they move at nearly the speed of light. Since nothing can move faster than light, it is very hard to move faster than a neutrino. Thus neutrinos, by and large, always seem to be left-handed for neutrinos and right-handed for antineutrinos. And, since angular momentum (i.e., spin) is conserved, if you have an interaction involving the weak force in which a neutrino is made in association with a charged lepton, the constraints on the spin of the neutrino restrict the possible spin of the charged lepton.

But the bottom line is not that there are restrictions on the spin of matter and antimatter, but rather that the weak force only interacts with a specific spin configuration for matter, and the opposite for antimatter. This suggests that it is possible that right-handed neutrinos could exist, but they have never been seen. This is because neither the strong nor electromagnetic force will interact with any form of neutrinos, and the weak force doesn't interact with right-handed neutrinos. The jury is still out on whether noninteracting right-handed neutrinos can exist.

One final caveat. Neutrinos are nearly massless but, since 1998, we've known that they have a small mass. Thus, it is possible, at least in principle, to go faster than them and thus change them from left-handed to right-handed. This requires scientists to invent slightly more complex measures of handedness, but that is beyond the scope of this book, and you can safely ignore this consideration unless you decide to take up the topic professionally.

Mirror, Mirror Redux

Theoretical physicists were astounded by the discovery that parity is not conserved in weak force interactions, but that's a pretty abstract thought. Practically, parity conservation means that if we start with a left-handed neutrino and swap left with right, we should get a right-handed neutrino. However, left-handed neutrinos have been observed, and right-handed ones haven't. And that one observational fact means that parity is not conserved in weak interactions.

There's a theorem in physics, called Noether's theorem (named after German mathematician Emmy Noether), that says that a conserved physical quantity means that there is a symmetry in the equations. Symmetry just means that you can make a change and not see any difference. If parity were conserved, that would mean that if you replaced the position variable x in the equations with $-x$, nothing changes. A simple example would be (equation) = x^2. If you make the replacement, you get $(-x)^2$, which is the same as the original equation. If parity were conserved, the equations would have to have this property.

Physicists use the letter P as shorthand for parity. In theoretical language, they'd say that P isn't a symmetry of the weak force. However, as soon as it was discovered that parity was not a symmetry seen in weak interactions, scientists came up with a modified symmetry.

If you play around with equations, you can make all sorts of changes and see if it makes a difference. One possibility is you can replace every example of matter in the equation with antimatter and vice versa. Scientists use the confusing term "charge conjugation" to describe this, and denote this with the letter C. But all they really mean is that swapping of matter and antimatter in the equations.

We can ask ourselves, what sort of practical prediction would a C symmetry mean? It means that if we start with a left-handed neutrino and swap matter and antimatter, this would lead to a left-handed antineutrino. However, such a particle is never seen, which means that C symmetry is also not a property of the weak force.

However, what about CP symmetry? What if we swap *both* left and right directions and swapped matter and antimatter? Well, again being practical, if we start with a left-handed neutrino and swapped left and right, we'd get a right-handed neutrino, which doesn't exist. But if we then also swapped matter and antimatter, we'd get a right-handed antineutrino, which does exist. Thus, physicists began to say that the weak force respects a CP symmetry, even though it doesn't respect C or P separately. This is demonstrated in Figure 6.5.

That's important, as CP symmetry means that matter and antimatter are the same, which is largely what we see in particle physics experiments. Of course, this doesn't help with the mystery of the preponderance of matter over antimatter in the universe. Finding a violation of CP conservation could be a clue in solving that perplexing problem. And that is the next story that we'll investigate.

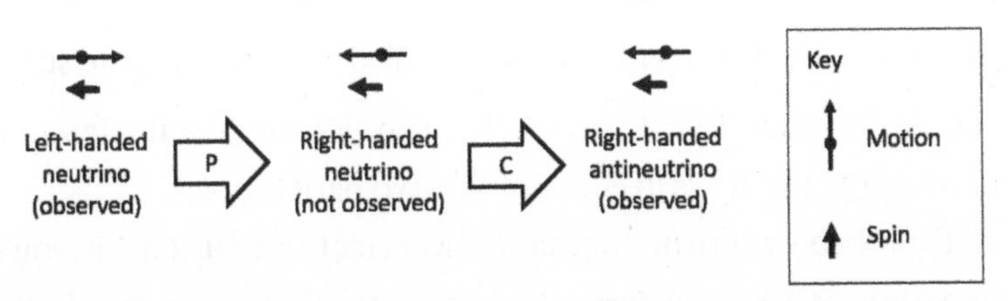

Figure 6.5 This figure shows that when one starts with a configuration that is observed in nature (a left-handed neutrino) and change the parity, the result is a configuration not observed in nature (a right-handed neutrino). However, if both parity and charge are changed, the final result (right-handed antineutrino) is observed. It doesn't matter whether one changes C and then P, or in the reverse order.

The tau-theta problem described in the previous section revolved around the particle that we now call the K^+ meson. The next story involves its neutral cousin—the K^0 meson.

Using modern nomenclature, the K^0 meson is a mixture of a down quark and a strange antiquark. Its antiquark equivalent is the $\bar{K}^0$ meson, which contains a strange quark and a down antiquark.

We saw in the last section that the K^+ meson didn't respect parity in weak force decays. The electrically neutral K^0 meson doesn't either. It can decay into either two or three pions, with a parity of either +1 or −1.

To understand the rest of this section requires invoking a little quantum mechanics, which I'm going to do only with a light touch. In quantum mechanics, little is certain, and the world is governed by probabilities. An electron can be in more places than one and a cat can be simultaneously alive and dead. (Actually, the cat thing is overdone and usually misunderstood, but that's a topic for another book.)

In the context of $\bar{K}^0$ particles, this means that the particle is simultaneously in both states of parity and, at the moment of decay, the universe "decides" that it will decay into one or the other state of parity. It's as if the K^0 meson literally consists of two different particles at the same time, one called K_S, with parity = +1, and the other called K_L, with parity = −1. You can then use the super simple equation $K^0 = F_{+1}K_S + F_{-1}K_L$. The "F's" are just the fraction of time that the K^0 is in either of the two parity forms.

It's probably worth noting that if you decide to read up on this, you'll find some books and papers refer to two other K mesons, specifically $K_1 (= K_S)$ and $K_2 (= K_L)$. There are reasons to do that, but I will stick with the L and S subscripts, for reasons that will become apparent.

OK, so that's the heavy quantum mechanical lifting. Hopefully it wasn't too bad. There will be a little more later, but we can rest for the moment. Now on to talking about a crucial experiment. The K^0 meson doesn't have a single parity, but K_S and K_L do. If we could somehow separate the two, we could look for violations of CP, which would be a big deal.

Remember that the reason we know that K mesons can have either parity is because of how it decays. If it decays into two pions, the decay state has positive parity. If it decays into three pions, the parity is negative. Now we need to think about energy.

If you zoom down a hill on a bike into a little valley and coast up the hill on the other side without peddling, you might be able to crest the second hill or not, depending on how high it is. It's easier to crest a short hill and hard to crest a high one. It takes more energy to go over the tall one. That's pretty intuitive.

In the decay of K^0 mesons, it can decay into two or three pions. Pions have a specific mass. The number doesn't matter for purposes of this conversation and different pions have slightly different masses, but we can ignore that here. And, because of Einstein's $E = mc^2$, we know that to make an object of mass m, you need a certain amount of energy, specifically E/c^2.

So, when a K^0 meson decays into two pions, it needs two pion mass units of energy, and if it decays into three pions, it needs three pion mass units. Just like cresting a bigger hill is harder than a shorter one, it's harder to decay into three pions than it is to decay into two. And, because it's harder, it takes longer to accomplish. We even have numbers we can associate with the decay of K mesons. The K_S decays in about 9×10^{-11} seconds, while the K_L decays in 5×10^{-8} seconds, or about 550 times longer. That's where the L and S come from—long versus short.

So we can use this difference in lifetime as the basis of an experiment. Suppose we took a beam of K^0 mesons and shot it off in some direction. The K_S component would decay away, leaving only the K_L component. Thus, in a K^0 beam, you'd expect the two-pion decays to be near the location where the K^0 was produced, and later, you'd expect to see only three-pion decays. Very clever.

This experiment was performed in 1964 by Jim Cronin, Val Fitch, and collaborators. The expectation if CP is conserved (and thus matter and antimatter are on equal footing) is that they would see only three-pion decay in distant detectors. However, when they ran the experiment, they found that for every 500 three-pion decays, they found 1 two-pion decay. And, just like Wu's experiment killed parity conservation for weak force interactions, the Cronin and Fitch experiment killed CP conservation. The universe does know the difference between matter and antimatter—at least a little bit.

The neutral K meson story isn't quite complete. If you think about it, I told you that K^0 mesons were a mix of the K_S and K_L states. However, after the beam had coasted a bit, all of the K_S mesons were gone, meaning that the beam was, strictly speaking, no longer a pure K^0 beam.

OK, we're going to jump back into some tricky quantum stuff. But we'll return to the bottom line at the end, so it's OK if you feel that you don't quite get all of the details.

I told you earlier that the K^0 meson is a mixture of a down quark and a strange antiquark, while the $\bar{K}^0$ meson contains a strange quark and a down antiquark. What I didn't tell you is that the weak force allows one to morph into the other and back again.

Figure 6.6 shows how that happens. Inside the K^0 meson, the down and antistrange quark swap a negative W boson, which I remind you is one of the carriers of the weak force. This changes the down quark into an up, charm, or top quark—we don't know

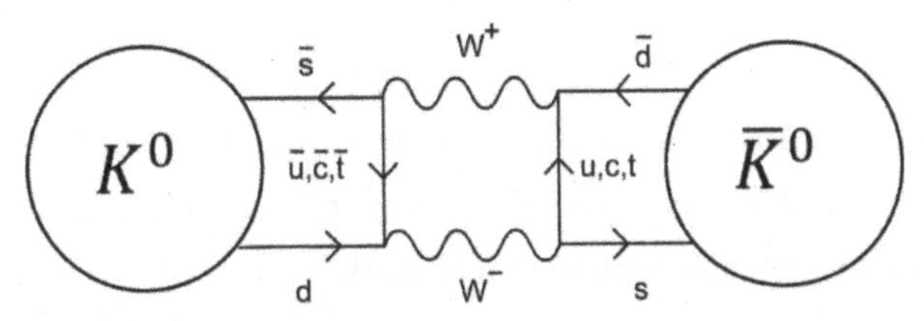

Figure 6.6 A neutral K meson (down and antimatter strange quark) can oscillate into an antimatter neutral K meson (strange and antimatter down quark). It does so by using W boson emission to change the quarks' identities

which, as any of them can happen. This also changes the strange antiquark into an up, charm, or top antiquark. This mixture of up, charm, and top quarks then emits a positive W boson, and now we have a strange quark and a down antiquark. Through this seemingly magical process, the world converted a K^0 meson into a $\bar{K}^0$ one.

And the process repeats itself. A beam of K^0 mesons oscillates into $\bar{K}^0$ mesons and back again. Matter becomes antimatter and then antimatter becomes matter.

Now we have to be careful, because inside every meson is a quark and antimatter quark, so we haven't quite turned matter into antimatter, but we've certainly swapped which quarks are matter and which are antimatter. This is a fascinating process and has been observed by looking at the beam when it decays. The decay into pions we've already discussed isn't the only decay, and other decays involving electrons can tell you whether the particle was a K^0 or $\bar{K}^0$ at the moment it decayed. Very cool.

OK, this is the moment of quantum magic. I told you that the K^0 meson was a quantum mechanical mix of K_S and K_L components and that's true of $\bar{K}^0$ mesons as well. That might have been almost obvious. But what isn't so obvious is that the converse is true. The K_L meson can be written as a quantum mechanical mix of K^0 and $\bar{K}^0$

mesons, as can the K_S meson. If you're a math-y kind of individual, you can write down the equation for the K^0 that I wrote a little while ago and a similar one for the $\bar{K}^0$ meson and solve it for the K_S and K_L ones. Or not. Sounds like a lot of work.

If you're not a math-y person, we can just think about the consequence. If the K^0 beam converted into a K_L beam, and a K_L beam is a mix of K^0 and $\bar{K}^0$ mesons, that means that when Cronin and Fitch saw the surprising two-pion decay, they were seeing decays of a mixture of matter and antimatter neutral K mesons.

Furthermore, subsequent measurements of the decay of K_L mesons into pions, electrons, and neutrinos gave a final clinch. If CP were conserved, we'd expect that when the long-lived neutral K meson decayed in this way, it would make neutrinos and antimatter neutrinos in equal quantities. However, decays into neutrinos were slightly favored. The universe does indeed seem to know the difference between matter and antimatter. And that provides us an opportunity to try to understand exactly how the universe became matter-dominated. And it was a Russian physicist that sketched out how.

The Sakharov Conditions

The universe is made of protons, neutrons, and electrons. Protons and neutrons are members of a class of particles called baryons, which just means that they contain three quarks. Baryons are much heavier than electrons, so if matter is favored over antimatter, the bulk of the extra matter is tied up in the baryons. Accordingly, the process whereby the matter excess seen in the cosmos is created is called *baryogenesis*, basically "the creation of baryons." This isn't quite fair, as the matter/

antimatter imbalance must also occur in electrons (i.e., leptons), but the term *baryogenesis* is more common, so we'll stick with it here.

Given the complexity of the problem of the asymmetry of matter and antimatter, it would be very helpful if someone could lay out in a clear and concise way what would be required for the symmetry between matter and antimatter that we see in particle accelerator experiments to become the asymmetry between matter and antimatter we see in the cosmos. As luck would have it, somebody has done that for us.

In 1967, Soviet physicist Andrei Sakharov wrote a paper enumerating three necessary conditions required for *baryogenesis*. They are as follows:

1. Baryon number violation: This means that a process must exist that favors baryons over antimatter baryons, producing more of the first than the second. If the universe originally made baryons and antibaryons in equal quantity, something that happened later must favor creating baryons more than antibaryons. This condition is pretty obvious.
2. C and CP violation: This means that both C and CP are not conserved. If there were a process that converted a balanced universe into some universe with extra baryons, if charge and charge/parity were conserved, then that means that the process converting a balanced universe into some universe with extra antimatter baryons would be the same. That's what the symmetry means—matter and antimatter processes must be identical. Thus, there must be something that favors matter over antimatter.
3. Thermal disequilibrium: This means that things are unchanging. Water held at 32° F (0° C) is a thermal equilibrium,

> because exactly as much water is melting as is freezing, leading to no change. Even at room temperature, water molecules are constantly bouncing into one another. Where one molecule gains energy, another loses it, and the net effect is no change. In the context of *baryogenesis,* if the universe is in thermal equilibrium, then whatever process is making an excess of baryons is exactly balanced with the opposite process that is taking extra baryons and going back to the balanced state. It's the freezing/melting thing all over again.

Sakharov's paper languished, unnoticed, for nearly fifteen years, until the era of grand unified theories arrived. His paper began to be appreciated around 1979. It's probably worth noting that Sakharov is more commonly known as one of the key contributors to the Soviet thermonuclear program, and he became a peace activist and Soviet dissident, eventually receiving the Nobel Peace Prize in 1975.

So what do we know about processes in the world that satisfy Sakharov's conditions? The standard model does not support any baryon number violation, meaning that new physics is required. It is thought that at much higher energies a process might exist with the mystifying name *sphaeleron process.* If this process exists, sphaelerons can convert baryons into antimatter leptons, or antimatter baryons into leptons. Using an illustrative (and technically wrong) example, it's as if a proton could convert into a positron (i.e., antimatter electron). Since before the transition there is one baryon (the proton), and after there is none, then this would change the number of baryons.

So that's a possibility, but it's only a hypothetical one. How about C and CP violation? Well two sections ago I described the Wu experiment, which demonstrated the violation of parity conservation in weak force interactions. That's because swapping the parity would convert a left-handed neutrino into a right-handed neutrino, and we don't see that in real life.

However, Wu's experiment also showed that C conservation was not allowed. The reasoning is the same—swapping matter and antimatter (i.e., C) would turn a left-handed neutrino into a left-handed antineutrino, which doesn't exist. So we know that the weak force violates C conservation.

Similarly, one section ago, we talked about the Cronin and Fitch experiment, which observed CP violation. Also, I didn't describe the extensive ongoing work in studying CP violation, which has been underway for decades. Experiments at both CERN and Fermilab have cemented the original conclusion that CP violation exists in the behavior of neutral kaons, and other experiments at the KEK laboratory in Japan, the Stanford Linear Accelerator Center in California, and Fermilab have demonstrated that mesons containing bottom quarks also exhibit similar behavior. (Neutral B mesons consist of a down quark and a bottom antimatter quark. The antimatter version of neutral B mesons contains the reverse.)

Accordingly, we see that the second Sakharov condition is at least observed in the standard model, although whether the amount of C and CP violation is the right amount is still not understood.

Finally, there is Sakharov's third condition, which was the existence of thermal disequilibrium. While the universe is pretty stable now, early in the history of the cosmos, the expansion of the universe was cooling it pretty quickly. Thus, early in the universe, there were moments where the energy was changing rapidly. Perhaps this is good enough to satisfy Sakharov's third condition.

However, with all that has been accomplished, we still cannot explain the matter and antimatter asymmetry. Ongoing experiments continue to try to find a variety of processes that would fit the bill. For example, at Fermilab, where I work, the staff is building the infrastructure to study neutrinos, which is a possible answer. The idea is the following.

My fellow scientists are upgrading the accelerator to increase the number of neutrinos that can be created per second. Fermilab already has the most intense neutrino beams in the world, but we need even more.

Once the upgraded beams are available, the plan is to shoot them in a westerly direction to a location over 1,300 kilometers away in the Black Hills of South Dakota. There lies an abandoned gold mine which is being turned into the world's premier neutrino laboratory. The mine is the Sanford Mine, and the laboratory is built in a cavern a mile underground called the Sanford Underground Research Facility, or SURF. In the cavern, scientist will install the Deep Underground Neutrino Experiment, or DUNE.

DUNE will study all manners of neutrino behaviors, but the most important for the question of the matter domination in the cosmos revolves around any possible differences in neutrino oscillation between neutrinos and antimatter neutrinos.

Neutrinos are unique in the panoply of fundamental particles of the standard model in that they "oscillate," which means that they change their identity. Over long distances, a beam of (for example) electron neutrinos will morph into muon and tau ones, before turning back into electron neutrinos and the process repeating itself. This has been well established since 2001 or so.

What Fermilab scientists will do is to shoot beams of first muon neutrinos and then muon antineutrinos to DUNE, trying to measure their oscillation processes. The most likely outcome is that the processes are the same for both neutrinos and antineutrinos. However, if they oscillate differently, this will violate both C and CP symmetries. That's an initial first step toward solving the problem of the observed matter excess in nature.

However, it's not enough. There are other considerations that go into the mix. For instance, if the *sphaeleron process* were true, then an excess of, for example, antimatter neutrinos (due to oscillation differences) would result in an excess of baryons, and that would give a way for Sakharov's first criterion to be satisfied. There's no guarantee that this will work, but observing a difference in the oscillation characteristics of neutrinos and antimatter neutrinos would be a first step. It is expected that the DUNE experiment will see first beam toward the end of the 2020s, although the schedule could slip. It always does.

The process that Fermilab scientists are hoping to find is called *leptogenesis*, which is equivalent to *baryogenesis*, if the *sphaeleron process* can convert antimatter leptons to baryons. Clearly there's a lot of work to do, and nobody should believe that they know the eventual outcome. But it's a promising prospect, and we'll know the answer in a decade or so.

Alternative Ideas

There are some people who think that the whole matter/antimatter asymmetry is much ado about nothing. While these are not universally accepted, what are other options?

A common proposal is that the observed preponderance of matter in the universe is illusory and that we've been totally misled. After all, we know that the Earth is made of matter, but what about other objects in the universe?

Well, we can be totally certain that the Moon and other planets are made of matter. After all, Neil Armstrong lived long enough to say, "The Eagle has landed." Had the Moon been made of antimatter, when the landing module touched down, it would have destroyed itself in a fireball that would have been visible from Earth. And all of the probes that have been landed on Mars, Venus, and the other planets guarantee that they are made of matter as well.

What about distant stars and galaxies? Could they be made of antimatter? No, they really can't be, either. How do we know?

Let's assume a star or galaxy is made of antimatter. If it were true, we couldn't tell the difference in our telescopes. An antimatter star would emit light, and that light is identical to the light emitted by our star. So no help there.

However, our star (and all stars) emits what we call the solar wind, which is a steady stream of protons and electrons. They wash over the Earth and cause the auroras seen by the North and South Poles.

Now the solar wind heads out into interstellar space, which is awash with hydrogen gas. We can see that gas using radio telescopes, which detect the radio emissions of hydrogen atoms and molecules. And, given that radio waves are a form of electromagnetic energy just like photons, we can't tell from our radio measurements whether the interstellar gas is matter or antimatter.

However, when matter and antimatter mix, we do know what happens there and we know very precisely. If a hydrogen atom hits an antimatter hydrogen atom, the proton and antimatter proton annihilate and release a lot of energy. That energy can come in a variety of forms which we could detect, but the details of the energy release can be a bit murky.

However, when an electron meets an antimatter electron, the situation is far more precise. An electron and positron annihilate and always release two gamma ray particles, each with an energy of exactly 0.511 million electronvolts (MeV). This is true even if the electron and positron are moving at commonly seen astronomical velocities.

Thus, if an antimatter star were surrounded by a matter cloud, or vice versa, hydrogen and antimatter hydrogen would run into one another, releasing a lot of energy and heating up the gas but, most importantly, releasing a whole bunch of 0.511 MeV gamma rays, which we could detect.

And the same thing is true of galaxies. Galaxies are also embedded in clouds of gas and, again, we see essentially no evidence of 0.511 MeV gamma rays. Accordingly, we can conclude that there are no (or, at most a very tiny number of) antimatter stars and galaxies out there. So we can put that conjecture to bed.

A similar hypothesis suggests that the separation of matter and antimatter exists on much broader scales. We know that the radius of the entire universe is at least 500 times larger than the visible universe, meaning that the volume of the entire universe is at least 125 million times bigger than the farthest that even a perfect telescope could see. Perhaps matter and antimatter are clumpy on huge distance scales—larger than tens of billions of light-years. What can we say about that?

Well, here we can say much less. However, it's hard to conceive how such large structures could form. Thus, most scientists put this aside as highly improbable and definitely untestable.

There is another school of thought about the cosmos's matter/antimatter asymmetry. These scientists essentially just kind of shrug their shoulders and say, "Well, perhaps the universe was just created with the asymmetry at the very beginning." If this were the case, all of the symmetries we see in the creation of matter and antimatter in our particle physics experiments are completely valid, but a tad misleading. And it would mean that all of our experiments that we're conducting and planning for the future are just totally misguided. They will always fail, because the laws of nature are actually totally symmetric.

That would be incredibly frustrating, as scientists would see the null outcome of experiment after experiment, and they would have to come up with theory after theory, each applying at a higher energy and with a smaller and smaller experimental signature. I guess it's job security, but it would surely get discouraging.

However, the hypothesis that the universe simply was created with an excess of matter doesn't help. For a theory of everything, we'd need to explain just why there was an initial excess. Thus, we'd be left with a very different intellectual direction to explore.

So let's recap. Where are we in the mystery of the missing antimatter? Our story begins with three important observations. First, energy converts into matter and antimatter in equal quantities in particle physics experiments. Second, shortly after the Big Bang, the universe was full of energy. And, finally, third—we now see a cosmos full of matter, with no antimatter. That's the mystery.

For the answer, we have a couple of options. We have the "because the universe was created with an asymmetry" solution, which requires that we somehow bake into the Big Bang theory the asymmetry at the moment of creation.

Then we have the other option, which is that the universe began with an equal amount of matter and antimatter, and some unknown physics somehow tipped the balance somewhat. We don't know if it happened in the grand unified era, say in the very first instants of creation, and at energies that are far, far, beyond what we can hope to generate in the laboratory for probably centuries, if not millennia. This would be a somewhat disappointing situation, but it's an entirely plausible one. After all, many of the proposed grand unified theories have all of Sakharov's three criteria baked into them. These theories often propose proton decay, which is clearly a baryon number-violating process. After all, before, you have one baryon (the proton) and after, you don't. That is a possibility, but testing the hypothesis is not in the cards anytime soon.

Then there is the other option, which is that the physics process that created the observed asymmetry between matter and antimatter occurred at an energy scale that is similar to the energy scale at which the Higgs field became relevant to the universe. This is what is called the electroweak symmetry breaking scale, and it is the energy at which the weak force became distinct from the electromagnetic force.

This energy scale is high, but we're beginning to explore it. We've discovered the W and Z bosons, as well as the Higgs boson, all of which are consequences of electroweak symmetry breaking. We've seen no real definitive evidence that this is the energy scale at which matter became more prevalent than antimatter, but certainly there have been many theoretical papers that explore the possibility. Ongoing experiments, like those at the Large Hadron Collider in Europe, are looking into the possibility. But there are also colliders of electrons and positrons at energies that make a neutral B^0 meson and antimatter $\overline{B}^0$ meson. These experiments are making precise measurements of CP violation in mesons involving bottom quarks.

And, of course, there is the future DUNE experiment, which has high hopes of detecting differences between the oscillation properties of neutrinos and antimatter neutrinos. They are following two ongoing experiments, one called NOVA and based at Fermilab, and the other called the T2K (Tsukuba to Kamioka) neutrino beamline in Japan. T2K and NOVA are already showing hints of CP violation in neutrino experiments, although the data are not yet precise enough to say anything definitive.

The bottom line is that, while we know that there is an unsolved question, we don't really know the answer, nor even which of many prospective solutions is correct. It's possible that the question could be answered by data accessible in the next decade or so, or it could be that we'll need data that will require energies that will not be achievable for perhaps centuries. Even more daunting is the possibility that the answer is baked into the very fabric of reality, which will require an even more imaginative answer, either a theoretical breakthrough or perhaps some idea that will take astronomical observations to resolve. No matter. It's a fine problem, and some very serious head scratching is in our future.

CHAPTER 7

ULTIMATE BUILDING BLOCKS

There are many mysteries in modern particle physics and each individual scientist might be more or less fascinated by any particular topic, and we've already encountered a few of them; but there is one that really bugs me. This quantum conundrum can be illustrated by a series of questions. For example, why do there appear to be unstable duplicates of the quarks and leptons? Why are some of them much heavier than others? Why are there three copies? Why not two? Or four? Or, even most satisfying, only one? What is the mechanism by which they convert one into another? There's a whole panoply of related questions that scientists variously call "the question of flavor" or "the question of generations."

Let's remind ourselves of the problem and let's start by looking at Figure 7.1. These are the quarks and leptons currently known to science. The left-most column contains all of the particles that make up the familiar cosmos. They are the up and down quarks, the electron, and electron neutrino. Up and down quarks are found inside protons and neutrons. Take those protons and neutrons and toss in the right number of electrons, and you have atoms. This, of course, leads to planets, galaxies, and us.

The electron neutrino isn't quite so familiar, but it plays a key role in a form of radioactivity called beta decay, which partially

Einstein's Unfinished Dream. Don Lincoln, Oxford University Press. © Oxford University Press 2023.
DOI: 10.1093/oso/9780197638033.003.0007

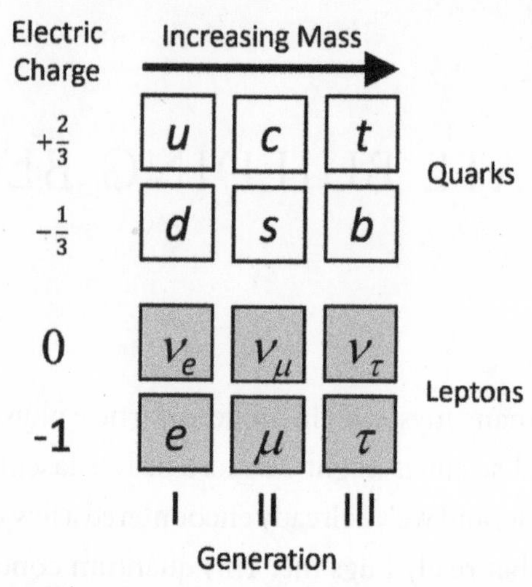

Figure 7.1 Known quarks and leptons, grouped by electric charge and with mass trends highlighted.

powers the Sun. So we are familiar with these subatomic ghosts, if only indirectly. Those are the inhabitants of the first column.

However, there is also the second column (charm, strange, muon, muon neutrino) and third (top, bottom, tau, tau neutrino). Each column is called a "generation," and just why more than one generation exists is unknown.

When a trained scientific eye, especially one with an interest in history, looks at Figure 7.1, they are struck by parallels to a similar situation in the mid-1800s. Perhaps lessons drawn from that earlier era might help us understand the origin of the recurring patterns seen in the particles of the standard model.

In 1869, Russian chemist Dmitri Mendeleev devised the organizing framework of modern chemistry. He created the first chemical periodic table of elements. By the mid-nineteenth century, the idea of atoms was already widely accepted, and the chemical properties of many elements had been determined. Mendeleev's creative contribution was to group elements with similar properties together, for example the highly reactive lithium, sodium, potassium, and cesium, or another highly reactive group of elements that include fluorine, chlorine, and bromine. While we now know of over one hundred elements, in the mid-nineteenth century not all elements had been discovered. In fact, Mendeleev knew of only fifty-six elements for which their mass had been determined, and a smattering of others for which the mass was not yet known.

He then wrote the name of the known elements on cards and arranged elements with similar chemical properties in rows, with the lightest on the left and the heaviest on the right. Modern versions of the table are organized into columns of similar elements, but the principle is the same.

Now let's put ourselves in the mindset of a top-notch scientist circa 1890 or so. The structure and patterns of the familiar and

H																	He
Li	Be											B	C	N	O	F	Ne
Na	Mg											Al	Si	P	S	Cl	Ar
K	Ca	Sc	Ti	V	Cr	Mn	Fe	Co	Ni	Cu	Zn	Ga	Ge	As	Se	Br	Kr
Rb	Sr	Y	Zr	Nb	Mo	Tc	Ru	Rh	Pd	Ag	Cd	In	Sn	Sb	Te	I	Xe
Cs	Ba	La	Hf	Ta	W	Re	Os	Ir	Pt	Au	Hg	Tl	Pb	Bi	Po	At	Rn
Fr	Ra	Ac	Rf	Ha	Sg												

Ce	Pr	Nd	Pm	Sm	Eu	Gd	Tb	Dy	Ho	Er	Tm	Yb	Lu
Th	Pa	U	Np	Pu	Am	Cm	Bk	Cf	Es	Fm	Md	No	Lr

Figure 7.2 Modern chemical periodic table of the elements. The gray elements are the ones known to Mendeleev in his original paper.

modern periodic table (columns of chemically similar elements, ordered in increasing mass, seen in Figure 7.2) were telling us something, but we didn't know what.

Any modern high school chemistry student now understands the patterns of the periodic table that puzzled the very best scientific minds of the late nineteenth century. Within each column, the outermost electrons (specifically those that can participate in chemical reactions) are in the same configuration. That's why they are all similarly reactive. And we also know that the atoms get heavier as one goes from top to bottom because the atoms contain an increasing number of protons and neutrons. What was mysterious a century ago is now explained.

The situation in the patterns found in the known quarks and leptons of the standard model, listed in Figure 7.1, is qualitatively similar to those found in the chemical periodic table of 1890—patterns without explanation. Let's review.

There are three quarks with electrical charge equal to +2/3 that of the proton. Similarly, there are three quarks with charge −1/3 that of a proton. And then there are three leptons, each with the opposite charge of a proton, and the three neutral neutrinos.

Generally, we organize these subatomic particles with particles of the same electric charge in rows. Furthermore, as we go from the left to right, the mass of the quarks and leptons increases. Another clue is that the particles in columns two and three are unstable and eventually decay into the familiar particles of the first generation. And, of course, only the first-generation particles are found in ordinary matter.

So that's what we know. Where do we go from here? Well, following the lessons of the chemical periodic table, it is at least plausible to consider the possibility that quarks and leptons (and perhaps the force-carrying particles as well) are composed of even smaller particles held within them. Under this hypothesis, the different quarks and leptons are simply different configurations of these hypothetical constituents. This is conceptually analogous to explaining the chemical reactivity similarities between, for example, sodium and potassium, as having the same configuration of electrons in the atoms' outer shell. The differences in chemical reactivity between, for example, sodium and neon, are explained on the two atomic species having different electron configurations.

The similarities between the three different generations have a couple possible explanations. The reason that sodium is heavier than lithium, which is heavier than hydrogen, is because the heavier atoms have more protons and neutrons in the nucleus. Perhaps the second- and third-generation quarks and leptons contain more constituents. At this point, this is only a conjecture.

On the other hand, there is another possible explanation for the higher mass of the second and third generations, and this explanation originates in Einstein's venerable $E = mc^2$. Under this hypothesis, possibly the top, charm, and up quarks all have the same constituents and in the same configuration. However, in the charm and top quarks, those constituents are orbiting one another with more and more energy. And since Einstein's equation says that energy and mass are equivalent, then perhaps the heavier mass of the quarks and leptons of the second and third generation is simply a reflection of that extra orbital energy. This is roughly analogous to when an electron in an atom jumps to a higher energy state. Extending this conjecture further, atomic electrons in a higher energy state eventually emit a photon and drop to the lowest energy state. Perhaps the decay of a charm quark into a down quark is a similar mechanism.

Mind you, everything I've said here is a mix of conjecture, hypothesizing, and fairly unapologetic guessing. Maybe the whole idea is wrong. Or maybe there's something to it. Let's take a look at some ideas that have been floated that explore this conjecture.

Preons

There is a school of thought that has postulated that quarks and leptons contain constituents, and those constituents have a name. That name is "preons." Now, before we dive into the preon idea, I should pause and tell you that this idea is not well regarded in the theoretical physics community. In fact, it's not popular at all. After I've described the idea in detail, I'll return to those objections, so you can have an informed opinion.

Personally, while I like the idea of preons very much, I do not believe in them. Nor should you, pending a confirming experiment. But they're neat. So let's dig in.

To begin with, the word "preon" is an overarching and generic term. There are actually dozens of preon models, and I'll only discuss one of them in detail and a second one to give a sense of the range of the possible. For each of these models, physicists have invented their own names for quark and lepton constituents. Some of the names include subquarks, maons, alphons, quinks, quips, rishons, tweedles, helons, haplons and Y-particles. Personally, I always kind of liked the terms quinks and tweedles. (I am agnostic on those particular models. I just like the names.) They have just the right level of silliness. However, for the remainder of this chapter, we'll go with the less colorful "preons."

The approach followed by most preon models is quite similar to the one followed by Murray Gell-Mann and George Zweig back in 1964, when they invented what we now call the quark model of mesons and baryons. They looked at the plethora of mesons and baryons that had been discovered in the 1950s and early 1960s and asked themselves what lessons the structures in the patterns seen in those particles were teaching us. They postulated the up, down, and strange quarks and then proposed that baryons (like the proton and neutron) were composed of three quarks, and the mesons (the pion being the most common) being composed of a quark and antimatter quark. The initial paper did not talk about the strong force, which governed the behavior of quarks and bound them together into baryon and mesons. That came later.

Perhaps the first paper released on preon theory was published in 1974. Written by Jogesh Pati and Nobel Prize winner Abdus Salam, it postulated a substantial modification to the standard model. The accepted theory claims that there are three distinct types of strong force charges—three "colors," called red, green, and blue. What Pati and Salam proposed was that there were actually four colors. The fourth color was what differentiated quarks and leptons; thus, the forth color could be called "lepton." Lepton was neutral in the sense of the original three colors, which makes sense, given that the leptons do not participate in the strong nuclear force. This particular preon model is interesting but is perhaps a little less intuitive than some of the others; thus, I mention it only due to historical interest.

Perhaps the simplest of the preon models was independently proposed in 1979 by Haim Harari and Michael Shupe, and later expanded by Harari and his student Nathan Seiberg in 1982.

This particular approach postulates that quarks, leptons, and the force-carrying particles are made of three preons. There were the (+) preons, with a charge of +1/3 and the (0) preons, with an electrical charge of 0. The antimatter charged preon ($\bar{0}$) had a charge of −1/3, while the electrically neutral antipreon can be denoted by the symbol $\bar{0}$.

In this model, the quarks and leptons can each be constructed by the combination of three preons. The electric charge of each subatomic particle then makes sense, for example the positron (i.e., antimatter electron), with its charge of +1, can be made from three + preons. Table 7.1 shows the preon content of the particles of the first generation.

Now the force-carrying particles, the charged W bosons, the Z boson, and the photon, are also made with preons. All of them except the photon are composed of six preons, while the photon is made of only two. Table 7.2 shows the preon content of the

Table 7.1 The Harari/Shupe preon content of the particles of the first generation of the standard model.

Charge	Preon Content	Known Particle
+1	+ + +	Positron
+2/3	+ + 0	Up quark
+1/3	+ 0 0	Down antiquark
0	0 0 0	Electron neutrino
0	$\bar{0}\,\bar{0}\,\bar{0}$	Electron antineutrino
−1/3	$-\bar{0}\,\bar{0}$	Down quark
−2/3	$--\bar{0}$	Up antiquark
−1	− − −	Electron

Table 7.2 The preon content of the bosons that govern the weak and electromagnetic forces.

Charge	Preon Content	Particle
+1	$+++000$	W^+
−1	$---\bar{0}\bar{0}\bar{0}$	W^-
0	$000\,\bar{0}\bar{0}\bar{0}$	Z (four versions)
	$+++---$	
	$++--0\,\bar{0}$	
	$+-00\bar{0}\bar{0}$	
0	$+-$	Photon

force-carrying particles. The gluons are a bit more complicated and omitted.

Ok, so that's great and all, but does it work? Let's just take an example of a well-known physics process and see if the preon approach agrees with what we know. Let's use beta decay as our example.

In beta decay, a neutron in a nucleus converts into a proton, electron, and electron antimatter neutrino. A semi-familiar beta decay involves carbon 14, which is used to date ancient materials that were once living, like wood, leather, bone, and so on. Carbon 14, which has six protons and eight neutrons, turns into nitrogen 14, which has seven protons and seven neutrons, along with the requisite electron (beta particle) and antineutrino ($^{14}_{6}C \rightarrow {}^{14}_{7}N + e^- + \bar{\nu}_e$).

Before looking at this reaction at the preon level, we need to understand what's happening with the quarks. A neutron consists of two down and one up quark, while a proton is two ups and one down. The only force that can change a quark's identity is the weak nuclear force, specifically with the emission of a W boson, in this case a negative W boson (W^-). The W^- then converts into an

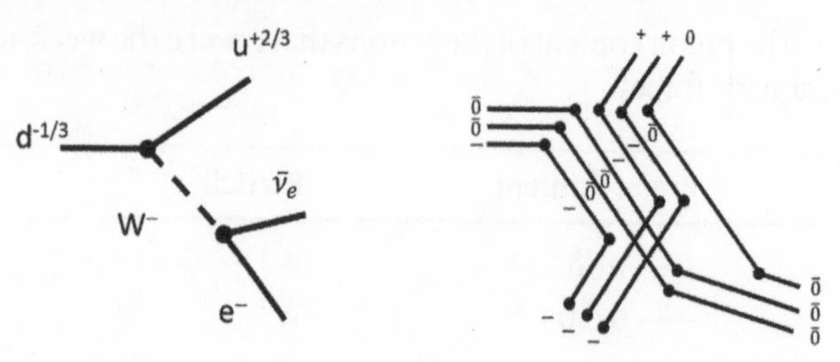

Figure 7.3 Beta decay at the quark level (left) and preon level (right).

electron and the antineutrino ($d^{-1/3} \rightarrow u^{+2/3} + W^{-} \rightarrow u^{+2/3} + e^{-} + \bar{\nu}_e$). This process is shown in Figure 7.3.

Figure 7.3 also shows the same process at the preon level. All of the preons of the down quark ($-\ \bar{0}\,\bar{0}$) flow into the W^{-} ($---\ \bar{0}\,\bar{0}\,\bar{0}$). As the W^{-} is being created, three pairs of matter/antimatter preons are made (+ –), (+ –), and ($0\,\bar{0}$). The matter preons of these pairs (+ + 0) form the up quark, while the antimatter preons ($--\ \bar{0}$) help create the W^{-} particle. Finally, the W^{-} particle decays, creating an electron (– – –) and an antimatter neutrino ($\bar{0}\,\bar{0}\,\bar{0}$). Thus we see that, at least in this particular preon model, we can use it to consistently model beta decay.

Let's take another simple subatomic process, the annihilation of an electron and positron, making a photon, which then decays into an up quark/antiquark pair ($e^{-}e^{+} \rightarrow \gamma \rightarrow u\bar{u}$). Figure 7.4 shows this, first at the quark level, and then at the preon level. An electron (– – –) and positron (+ + +) merge. Two pairs of positive (+) and negative (–) preons merge and annihilate, leaving the third pair to make the photon (+ –). Then, when the photon decays, the (+) preon goes to the up quark, while the (–) goes to the antimatter up quark. Simultaneously, a (+ –) and ($0\bar{0}$) set of preon/antipreon pairs are made. The matter preons flow into the up quark, and the antimatter ones flow into the antimatter up quark. Again, we see

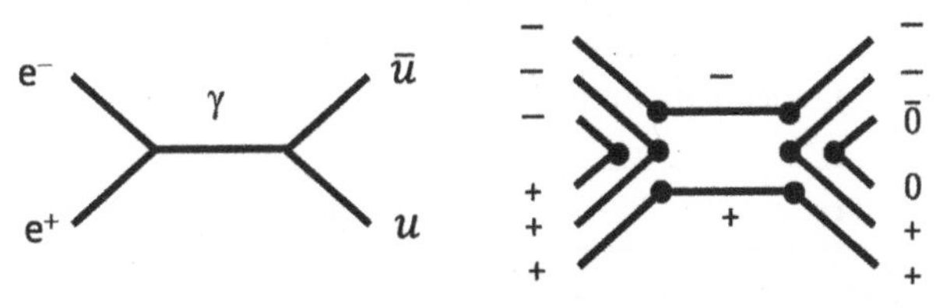

Figure 7.4 The $e^+e^- \rightarrow u\bar{u}$ process at both the quark/lepton (left) and preon (right) level.

how this particular preon conjecture can consistently model a well-known subatomic process.

While the Harari/Shupe model of preons is simple and easy to understand, it is not unique. German physicists Harald Fritzsch and Gerd Mandelbaum proposed an entirely different preon model in 1981. In their model, they proposed four preons, which are listed in Table 7.3.

Their preons are combined in pairs to make the particles of the standard model, as shown in Table 7.4. However, in their model, the gluons and photons have no preon content. Like in the standard model, in the Fritzsch/Mandelbaum preon model, these particles have no internal structure.

We see that the Fritzsch/Mandelbaum model is extremely different from the Harari/Shupe one, and the data do not favor one of them over the other. (Indeed, we will return to criticisms of preon models in general in the next section.)

You should realize that the preon models, as we've discussed them so far, are not actual physics theories. They're more like state-of-the-art quark theory in 1964. It's like the symmetries discussed in Chapter 3.

Compare that to the theory of the strong nuclear force, which describes in detail how the gluons bind together quarks into mesons and baryons. So far in our discussion, we're nowhere near that level of sophistication.

Table 7.3 The preons found in the Fritzsch/Mandelbaum model.

Preon	Electric Charge
α	+1/2
β	−1/2
x	+1/6
y	−1/2

Table 7.4 Preon mix of particles of the standard model in the Fritzsch/Mandelbaum preon model.

Particle	Preon Content	Particle	Preon Content
Electron neutrino	(αy)	W^+	$(\alpha\bar{\beta})$
Electron	(βy)	W^-	$(\bar{\alpha}\beta)$
Up	(αx)	Z	Mix of $(\alpha\bar{\alpha}+\beta\bar{\beta})$
Down	(βx)	Higgs boson	Mix of $(\alpha\bar{\alpha}+\beta\bar{\beta})$

While there are dozens of preon models, all different in one way or another, they all have to postulate some mechanism that binds the preons together into the quarks and leptons. Different theories postulate different forces—indeed, as we see from the differences in the Harari/Shupe and Fritzsch/Mandelbaum models, the different theories must be different in detail. However, just as the word "preon" has evolved to be a generic word that can be used for all quark and lepton constituents, the term "hypercolor" has evolved for the charge that holds preons together. This is roughly analogous to the electric charge of electromagnetism, or the color charge of the strong nuclear force. Indeed, the term "hypercolor" is a nod to the fact that the force must be stronger than the strong force.

With the introduction of the preon model, let's now turn our attention to the many reasons why the theoretical community is disenchanted with the idea.

Preon Criticisms

There are many reasons why preon theory is not well regarded. One criticism is simple: there is no experimental data confirming them. That's not reason to reject the idea entirely—after all, it took five decades to find the Higgs boson—but it's always valuable to remember that physics is ultimately an empirical endeavor and we should believe nothing until we've directly detected it.

However, there are more general criticisms which revolve around conceptual tensions between the size of quarks and their mass. Explaining this requires some of the lessons of 1920s quantum mechanics.

First, let's start with putting some numbers on what existing data tell us about the size and mass of preons. And, before that, let's talk about the size of well-known subatomic objects. The radius of an atomic nucleus is about 10^{-14} meters, while the radius of a proton or neutron is about 10^{-15} meters. Quarks and leptons are much smaller, with a radius smaller than about 5×10^{-20} meters.

I said "smaller than" because we actually don't know if quarks and leptons have any size at all. Indeed, in the standard model, quarks and leptons have no size, beyond the wave/particle duality of quantum mechanics. And the reason we can set limits on the size of quarks and leptons is that we've looked. We know the smallest objects we can resolve with the most powerful particle accelerators and sophisticated detectors and that size is about 5×10^{-20} meters. Since we see no evidence indicating that quarks and leptons have a size, we can only say that if they do have a size, it's smaller than that.

You might wonder how we can measure the size of such small objects. Actually, there are two different, but interrelated methodologies. The basic idea boils down to this. In quantum mechanics, all subatomic objects have both a particle and wave nature. Electrons, photons, quarks, and so on have an associated wavelength, and that wavelength depends on the energy carried by the particle. (Technically, it's the momentum, but at high energies energy and momentum are the same, so I'm sticking with the more familiar term.)

When one is irradiating any object with a wave of some sort, the wave only really interacts with the object if the wavelength is smaller than the object. If the wavelength is larger than the object, it just sort of rolls over it without even noticing it's there. A familiar example might be ocean waves washing over the piling holding up a pier. If the water waves are long, basically they don't see the piling. If the wavelength of the water waves is short, the piling disturbs the wave pattern, and you can see it.

The wavelength of subatomic particles is inversely related to the object's momentum, or energy. The absolutely highest possible energy particle that could conceivably be produced in the Large Hadron Collider (LHC) is one that is 6.5 trillion electron volts. That's the highest. The more reasonable and common "highest energy" is more like 0.1 trillion electron volts. If you calculate the wavelength associated with these high-energy particles, you find them to be about 2×10^{-19} meters for the high-energy example, and 10^{-17} meters for the low-energy one. That's, ballpark, between 1/80 and 1/5,000 the size of a proton.

This sets the range of the size of objects one can directly image in LHC collisions. The bigger size is pretty common, while the smaller size is vanishingly rare. When you balance how often collisions involving energies in that range occur, you might say that the LHC has substantial data looking at objects that are about 10^{-18} meters, with some data in the 10^{-19} meter range and a tiny bit smaller still. It's impossible for the LHC to see objects much smaller than that.

I said that the size of quarks and leptons must be smaller than that (5×10^{-20} meters), but that's using more sophisticated data analysis techniques, beyond the scope of this book, which squeeze every bit of performance out of the equipment.

Scientists also look for deviations from predictions of the standard model and in how often quarks decay without changing their charge (for example, top→charm, which is never seen, as compared to top→bottom, which is very common). The first is a decay from a quark with + 2/3 charge to another with the same charge, while the second is a decay from a +2/3 charge quark to a −1/3 one.

The point here is that if you use the well-tested standard model, you can set even smaller limits on the size of quarks and leptons, but even basic quantum mechanics sets stringent limits on their size, requiring that they be smaller than 1/5,000 times the size of a proton.

Now, if quarks and leptons are that size, it stands to reason that preons, if they exist, must be smaller. What about masses?

Well, we have tons of experience converting energy into mass, under the auspices of Einstein's familiar $E = mc^2$. We've found all the quarks and leptons, including the massive top quark, with a mass of 172 billion electron volts, over 180 times heavier than the proton. And the LHC can generate energies as high as 13,000 billion electron volts, meaning that it could create even heavier particles. (And 13,000 is the absolute maximum. Realistically, the LHC could create particles in the 2,000 billion electron volt range, give or take.)

Since we haven't created anything that looks like preons, we can conclude that, if they exist, they have a mass higher than in the ballpark of 2,000 billion electron volts. That's not so ridiculous. Could be.

However, we need to remember that up and down quarks have a mass of about 0.003 and 0.005 billion electron volts, respectively. Electrons are lighter still, with a mass of 0.0005 billion electron volts. And, of course, photons are massless. This, of course, poses a problem. Using the Harari/Shupe model, we need three preons to make a quark or lepton or two preons to make a photon. That raises the question of how we take two particles with a mass of over 1,000 billion electron volts each and combine them to make a photon with a mass of zero. Or, in the case of the up quark, how can we take the three preons, with a combined mass of at least 2,000 billion electron volts or more, and get a final particle with a mass of 0.003 billion electron volts. Sounds like a problem, huh?

Well, yes and no. Obviously, if preons exist, they form what is called a "bound" state in a quark, for example. And binding energy is negative. To see that, think about a simple hydrogen atom, with one proton and one electron, each with their own mass. If they are combined into a hydrogen atom and we want to pull them apart to be two separate and noninteracting particles, we have to add energy. That means that the hydrogen atom has less energy (and therefore less mass) than a proton and electron separately. Thus, binding energy must be negative.

Now suppose the same thing is true with preons. Suppose the three heavy preons in the up quark are held together with a ginormous amount of negative binding energy. If the negative binding energy is just a tiny bit less than the mass of the three preons, then that would explain it.

That is possible, but we'd have to figure out why three preons bind together to make quarks (mass about 0.003 or 0.005 billion electron volts), electrons and positrons (mass of 0.0005 in the same units), and neutrinos (with a mass of less than 0.000000001 in those units). Furthermore, why just those three classes of masses and not others?

Well, perhaps there are others. There are, after all, the other, heavier quarks and leptons. But you'll note that in the introduction of the preon models, they only included the quarks and leptons of the first generation. How do we distinguish those from those of the second and third generation? After all, the motivation to think about preons came in part from the regularity of the generations. There is no single unique proposed answer to that question.

Then there is the question of the Higgs field and how it plays into preon theory. As it stands in the standard model, the mass of the quarks and charged leptons (and maybe neutrinos) is generated by their interaction with the Higgs field. In order for the preon model to hold in its simplest case, we'd have to figure out a way for the simple preon configuration that generates, for example, the up quark to interact with the Higgs field in different ways to make the charm and top quark. That's not impossible, but it runs afoul of the very successful Higgs theory of the standard model.

And this ignores the fact that the Higgs boson of the standard model is a fundamental (e.g., structureless) object, yet in some preon models (e.g., Fritzsch/Mandelbaum), the Higgs boson is itself composite. Even more complicated, there are other theories involving a concept called technicolor that also postulates a substructure to the Higgs boson. That concept is beyond the scope of this book, but it adds to the confusion.

There are criticisms of preon models that are more technical, having to do with a variety of technical mathematical symmetries. These criticisms are definitely beyond this book, which is simply a survey of the range of possible explanations of the patterns seen in the quark and lepton families.

In any event, the preon model is not a popular one in the theoretical physics community and is not being actively pursued for many reasons. Even though it has some attractive features at first blush, other ideas have led to greater theoretical interest. Let's explore another interesting proposal.

Leptoquarks

When one looks at the particles of the standard model, in addition to the obvious patterns that led to preon models, there are another series of interesting questions. Why are there quarks and leptons? Why are there two classes of particles? Why do leptons have integer charge, while quarks have fractional charge? Why is there a difference of one unit of electrical charge between quarks within a generation and one unit of charge between the leptons? Is that significant? Given that quarks and leptons can be arranged in families, is there a deeper connection between quarks and leptons that we have not discovered? In short, rather than wondering what lessons the patterns in the three quark and lepton generations are telling us, what lessons are the similarities and differences in quarks and leptons telling us?

We had one possible explanation about the fractional versus unit charge between quarks and leptons that we talked about in the section on preons. However, there is another idea that scientists have explored in trying to understand the mysteries of quarks and leptons and that is to postulate an idea called "leptoquarks."

In Plato's "Symposium," the character Aristophanes explains his theory of the origins of romantic love. In the beginning, people consisted of two heads and eight limbs. Two people were blended into one, either a man and a woman or two men or two women. Following an attack on the gods, Zeus split people into two, and thus men and women were formed. According to him, we've been looking for our other halves ever since.

In a similar albeit poetical vein, perhaps there is a similar phenomenon going on in particle physics. How would that go? Well, let's start by talking about the important properties of quarks and leptons individually.

Quarks are fermions. This means they have a quantum mechanical spin of ±1/2. They have fractional charge (+2/3 for up-type and −1/3 for down-type). They carry the strong nuclear charge of red, blue, or green. They also carry the weak charge. There are matter and antimatter equivalents of both. They have a baryon number of +1/3. This is because baryons like protons and neutrons have a baryon number of +1, and there are three quarks inside baryons. Because they are not leptons, they have a lepton number of 0. Those are the quarks.

Leptons are also fermions, with spin of ±1/2. They have integer charge (−1 for electron-type and 0 for neutrino-type). They have no strong nuclear charge. They do carry the weak charge, and there are matter/antimatter equivalents for both. Leptons have a lepton number of + 1, while they have a baryon number of 0. Those are the leptons.

Now suppose there is a heavy particle, called a leptoquark, which has properties of both quarks and leptons. Leptoquarks have both lepton and baryon numbers. They have both a weak charge and a strong force charge. There are a great variety of electric charge they can have, for example an up quark (charge = +2/3) and an electron (charge = −1) will combine to have a charge of −1/3. Because all leptons have integer charge (0 or ±1) and quarks have a fractional charge of (±1/3 or ±2/3), leptoquarks all have an electric charge of the form (number)/3, where (number) is not divisible by 3. I will revisit electric charge in a moment.

Some models of leptoquarks act as if they were a combination of a quark and a lepton, with anti-leptoquarks appearing to be a combination of antiquarks and antimatter leptons. Other models predict leptoquarks that are a mix of matter and antimatter, having the properties of a quark and antimatter lepton, or lepton and antimatter quark.

Furthermore, there is the question of subatomic spin to consider. Quarks and leptons either have a spin of +1/2 or −1/2. Spins add, so a quark and lepton together have either a spin of 0 or 1 (or −1, but that's effectively the same as 1). That means that leptoquarks either have a spin of 0 or 1. That's an integer spin, which means that leptoquarks have the same kind of spin as the familiar, force-carrying particles, that is, photon, gluon, and so on.

Now for the tricky part. Do leptoquarks obey the rules of the weak nuclear force or not? The weak nuclear force only interacts with quarks and leptons with spin −1/2 and antimatter quarks and leptons with spin of +1/2. (You may have heard of right-handed or left-handed particles. This is the same thing, simplified.) The strong and electromagnetic forces don't care much about spin—they'll interact with matter and antimatter quarks and leptons with either spin.

So, if leptoquarks respect the rules of the weak force, only certain types of quarks and leptons can be created when leptoquarks decay. If leptoquarks don't respect the weak force rules, more options are available. And, when one combines the weak force rules with considerations like whether leptoquarks have spin of 0 or 1, and whether leptoquarks can act like they contain two matter or antimatter quarks or leptons, or whether leptoquarks act like they have a mix of matter and antimatter particles within them, this can restrict the possible kinds of leptoquarks that can exist.

Now that may all seem to have been a bit complicated—and it is—but it's the sort of thing you have to consider when you're trying to study the data to see if leptopquarks exist or not. Welcome to my world.

So let's return to the topic of electric charges. If one lists all conceivable charges of leptoquarks could possibly have, they are +5/3, +2/3, −1/3, and −4/3 (and, of course, the negative of these).

However, as I tried to warn you above, that's only if there are no restrictions on whether leptoquarks are all matter, or if they are a mix of matter and antimatter. Then there's the question of their spin and the second question of whether they follow the rules of the weak force or not. It's all a bit of a mess.

When studying the data to see if leptoquarks can be found, you have to make some assumptions about all of these properties. If you find leptoquarks that follow your assumptions, great. Collect your Nobel Prize. If you don't, then make a different assumption and do a different analysis. And, if that doesn't get your ticket to Sweden, you go back and try another configuration and try again, until all of the possible configurations have been tested. If you haven't been successful, you might think that you've ruled out leptoquarks, which would still be a scientific advance, as a lot of science is figuring out what things don't work. Knowing which ideas are wrong at least tells you where not to look.

However, all a failure to find leptoquarks tells you is that you didn't find them. It could be that they don't exist, or it could be that they are heavy enough that your accelerator doesn't have enough energy to make them. In that case, you build a more powerful accelerator and start over. Again . . . welcome to my world.

As complex as leptoquark theory is, it really isn't worth it to talk about every single possibility. But if we restrict ourselves to just one basic configuration, we can at least get a sense of what scientists look for. Take, for example, leptoquarks that contain only a matter quark and lepton, and furthermore, this particular version of leptoquarks follows the weak force rules. If that's the case, the possible leptoquark charges are +2/3, −1/3, and −4/3, and Table 7.5 lists them and the quarks and leptons into which they can decay. (The antimatter leptoquarks are the opposite charge and decay into antimatter quarks and leptons.)

It is thought that if leptoquarks exist, they are created in matter/antimatter pairs, just like other subatomic particles. So we see that the signature of finding leptoquarks in any particular particle accelerator collision is either the simultaneous production of a quark and antimatter quark and an electron and antimatter electron $(LQ \rightarrow q + e^-, \overline{LQ} \rightarrow \overline{q} + e^+)$.

An eagle-eyed reader might notice that the decay products of the leptoquarks of Table 7.5 are all members of the first particle generation. That's because many leptoquark models respect the same generational structure as quarks and leptons. Thus, a first-generation leptoquark will decay into up, down, electron, and electron neutrino of the first generation. A second generation leptoquark decays into a charm, strange, muon, and a muon neutrino, while a third generation leptoquark will decay into a top, bottom, tau, and tau neutrino. This isn't, strictly speaking, necessary, but most leptoquark models make this assumption.

It should also be noted that leptoquark and preon models can easily coexist. As seen in the previous section, the Harari/Shupe preon model says that an electron consists of three (–) preons, while

Table 7.5 This table shows the possible electric charge of leptoquarks and the quarks and leptons into which they could decay. Only shown are leptoquarks that contain matter both quarks and leptons, and which follow the rules for weak force reactions. The respective anti-leptoquarks have the opposite electric charge and they decay into antimatter quarks and leptons.

Electric Charge	Possible Decays
+2/3	$\nu_e + u$
–1/3	$e^- + u$ $\nu_e + d$
–4/3	$e^- + d$

an up quark consists of two (+) preons and one (0). Thus, a first-generation leptoquark with a charge of −1/3 would have a preon content of (− − − + + 0). Preon models don't require leptoquarks, and leptoquarks could exist, even if preons don't. But, should both exist, this would not pose a hardship for either model.

So what about the data? Is there any evidence for the existence of leptoquarks? Well . . . no. However, there was a bit of excitement back in February 1997, when a thrill ran through the physics community. At the time, the DESY laboratory in Hamburg, Germany, hosted what was called the HERA particle accelerator. This device collided protons and electrons at high energy. Two detectors using the HERA facility saw an excess of violent collisions, which resulted in very high-energy electrons and quarks being produced. Nobody could explain the result, but many scientists hoped that leptoquarks had been found. It was an exciting and giddy time.

Unfortunately, the excitement was short-lived. By May of that year, scientists operating at Fermi National Accelerator Laboratory, located just outside Chicago in the United States looked for the same signature that the HERA experiments reported and found nothing. The result turned out to be much ado about nothing, and the scientific world returned to their equipment, looking for more discoveries.

As of this writing (Fall 2022), many additional searches for leptoquarks have been performed, with none observed. The most stringent limits on leptoquark production come from data taken at the LHC, located in Switzerland. If leptoquarks do exist, they are very massive—over a thousand times heavier than a proton.

While no experiment can rule out leptoquarks completely—after all, they might exist, but be just a tiny bit heavier than your experiment can produce—the scientific community has relegated the leptoquark idea as one of many that has not yet been validated and may turn out to have simply been wrong. Such is the nature of scientific inquiry.

Superstrings Redux

Superstring theory was covered in Chapter 3, but it obviously has some relevance in any conversation about possible quark and lepton substructure. In superstring theory, the known quarks, leptons, and force-carrying particles are all just different vibrational patterns of a single one-dimensional string, vibrating in a ten- or eleven-dimensional space.

This hypothesis could explain the patterns seen in the known subatomic particles, thus solving the mystery of quark and lepton substructure. However, there remains the issue that the energy scale that we've explored in our current experiments is far removed from the energy scale at which superstring theory is supposed to reign—specifically, the Planck energy is about a quadrillion times higher than the energy that can be created by the LHC. This corresponds directly to the physical size range separating superstrings and quarks.

With such a large range in size and energy, it isn't ridiculous to imagine that there might be additional layers of matter between the known and the Planck scale. While acknowledging that preons may well not exist, we can use the term to stand in for any hypothetical building blocks of quarks and leptons. Loosely using the term, perhaps not only preons exist, but pre-preons, pre-pre-preons, and so forth. Indeed, it would be entirely reasonable to imagine that as we work toward discovering a theory of everything, that we might well find layer after layer of structure—things that we can't yet even imagine.

Another interesting feature of superstring theory is that in order for it to be mathematically consistent, the universe must have more dimensions than the familiar three of space and one of time. The dimensions we experience day to day all seem to be infinite in scale, with no end. In superstring theory, the extra six dimensions differ in a very important way. Those extra dimensions are tiny; they are of a similar size to the Planck length.

But this raises another point. If extra dimensions exist, it could be that not all of them are of Planck length. There could be extra dimensions that are a million times smaller than the 10^{-19} meter scale we can access with the LHC. A dimension of this scale is much smaller than we can study now, but much larger than the Planck scale. Although we have no reason to believe such intermediate dimensions exist, neither can our current data rule that out.

In short, superstring theory has been popular in the theoretical community, and it is a possible explanation for the patterns we see in the known quarks and leptons, but it is entirely theoretical in nature. Even worse from an experimentalist's point of view, the theory has not made a single testable prediction. There's simply no reason to accept the hypothesis, no matter how mathematically compelling it may be.

Returning to Reality

I've spent the last few sections talking about a series of theoretical ideas that might explain the patterns observed in the known quarks and leptons and could possibly lead to the discovery of yet another, and smaller, set of particles. However, theoretical conjecture is only as good as the experiment that confirms it. Are there any other hints we might draw from known data? I can think of two.

The first arises from the observation that there are three quark and lepton generations. Each generation seems to have the identical structure, and each subsequent generation is heavier than the one before. This leads us to an obvious question. Why are there three generations? Are there only three generations? Could scientists discover a fourth one?

The answer to that last question is "yes," at least in principle. However, there are a couple of reasons to believe that there are exactly three generations and no fourth. The first of these reasons comes from an accelerator-based experiment conducted in the 1990s.

It was in the 1960s that electroweak theory predicted the existence of the W and Z bosons as the particles that transmit the weak nuclear force. And it was in 1983 that they were discovered in a particle accelerator based at the CERN laboratory in Europe. Beams of protons and antimatter protons were collided together at high energy and that energy was transformed into W and Z bosons. These particles are incredibly massive, with the W and Z bosons having a mass of 80.4 and 91.2 billion electron volts, respectively.

With the discovery of these particles, the scientific community wanted to study them in much greater detail. The way to do that is to cause a beam of electrons to collide with a beam of positrons, with their energies carefully selected to be an energy that enhanced the production of these heavy particles. And so the Large Electron Positron (LEP) collider was built at CERN. It began operations in 1989, and it ran until 2000.

LEP was a large ring, 27 kilometers (16 miles) in circumference. Initially, it accelerated an electron and positron beam, each with an energy of 45.6 billion electron volts. They collided with a combined energy of 91.2 billion electron volts—exactly the energy needed to produce Z bosons. Over the course of operations, over 18 million Z bosons were produced, allowing the particle's properties to be measured in exquisite detail.

The Z boson interacts with particles that carry the weak force charge, which is true of all of the quarks, charged leptons, and neutrinos of the standard model. That means it can decay into any of the quarks and leptons.

The standard model makes very specific predictions about the fraction of time that the Z boson can decay into quarks (69.2%), charged leptons (10.2%), and neutrinos (20.5%). These predictions rely on there being three generations.

Now, the decay of Z bosons to neutrinos cannot be directly observed—after all, neutrinos pass undetected through the kinds of detectors used at the LEP accelerator. So,we can't see them. However, we can observe when the Z boson decays into charged leptons and quarks, allowing us to validate the predictions of the standard model of the decay of the Z boson into non-neutrinos.

If there existed more neutrinos than the known three, then the fraction of the time that Z bosons could decay into neutrinos would be larger than the prediction of the standard model. Similarly, if there were fewer neutrinos than the known three, then the Z boson would decay more often into the charged leptons and quarks.

Thus, by looking at the rate at which the Z boson decayed into the non-neutrinos, the LEP experiments could determine exactly how many generations of neutrinos exist, and they found that it was 2.984 ± 0.008. Since the number of generations must be an integer, this means that the LEP scientists proved that there were exactly three different types of neutrinos. This is one of my favorite physics results from the LEP accelerator.

The fact that three neutrino generations exist has also been demonstrated in measurements of the cosmic microwave radiation, which is the fossil remnant of the fireball of the Big Bang. That particular measurement is very interesting, but for purposes of this discussion it matters simply because it confirms the LEP result. There are three varieties of light neutrinos.

Given that there are exactly three varieties of neutrinos and each particle generation contains one neutrino, this then suggests that there are exactly three generations of quarks and leptons—not four, and certainly not more. The existence of three and only three generations is a very important clue, with important consequences. For example, in preon theory, each generation can be thought of as being the same as the first generation, but with the preons of subsequent generations simply orbiting faster. It's hard to understand why preons would have only three configurations available to them.

On the other hand, it is possible in some versions of superstring theory to predict that exactly three generations exist. This is a small triumph for superstring theory, but not a compelling one, as there are many versions of superstring theory and not all make this prediction.

It's worth noting that the LEP measurement doesn't, strictly speaking, say that there are only three versions of neutrinos. It simply says that there are three neutrinos into which Z bosons can decay. If a hypothetical fourth neutrino existed and had a mass higher than 45.6 billion electron volts, it would be so heavy that the Z boson couldn't decay into them. So it's possible that a fourth neutrino could exist, but it would be much heavier than the other three. This scenario is not excluded, but it certainly does seem that the data favor the conclusion that only three generations exist.

There is another important clue that experiments have given scientists fascinated by the patterns found in the quarks and leptons of the standard model. Of the known subatomic forces, only the weak force can transform one quark into another quark, and the same is true for leptons. If only the strong and electromagnetic force existed, then a top quark could never decay. However, a top quark can emit a W boson and convert into a bottom quark (i.e., $t^{+2/3} \rightarrow W^{+} + b^{-1/3}$). All other decays of quarks and leptons of the heavier generations into the lighter ones proceed via the weak force, through the emission of a W boson.

Thus, if anyone wants to understand the perplexing and pesky patterns found in the quarks and leptons, it's worth understanding the weak force very well. There is something about that force that the others don't have, and it is tied somehow into the generational structure of the known subatomic particles.

Like the other outstanding mysteries mentioned in earlier chapters, the fact that there exist quarks and leptons and that they are different is not explained. Nor do we understand why there are three duplicates of the quarks and leptons, when only one is required to build the matter of the universe. This is yet another mystery for a budding young scientist to solve and gain scientific immortality. Thus far, it has stymied me, but perhaps you will be the one to figure it all out.

CHAPTER 8

THE FUTURE

In the last seven chapters, I hope I have persuaded you that the search for the theory of everything is perhaps the ultimate goal of science. Questions about the origins of the universe, the nature of reality, and why the laws of nature are what they are have puzzled people for as long as we have historical records, and I strongly suspect long before that.

And we've made a lot of progress. As we learned in Chapter 2, we've devised the standard model, which describes the microcosm—the world of molecules, atoms, and even smaller objects. We've also created a very effective theory of gravity—one which predicts black holes, gravitational waves, and the expansion of the universe itself. We've even learned that space and time are two sides of the same coin.

But understanding our future prospects of a theory of everything isn't really about where we are now. It's a combination both of how we got here and what a path forward might look like. And on that, the name of the game is unification—showing that seemingly disparate things have common origins. We've encountered several examples in this book, but it's worth reminding ourselves what that means. As an example, consider two seemingly unrelated facts, for example the spike in radiation seen in ice cores

Einstein's Unfinished Dream. Don Lincoln, Oxford University Press. © Oxford University Press 2023.
DOI: 10.1093/oso/9780197638033.003.0008

drilled in the Arctic that began in the middle of the twentieth century and the extinction of many species of large mammals in dating from ten or twenty thousand years ago. They seem to have nothing in common.

And yet they're strongly connected. Humans hunted large mammals for food, and we detonated the first atomic weapon in the final year of World War II. Two very different observations had a common cause: us.

The past and the future of the search for a theory of everything are built from many similar sorts of connections. For example, we learned in Chapter 2 about how the behavior of objects falling on the surface of the Earth and the march of the planets across the sky could both be explained by what we now call gravity. Newton unified the Earth and the heavens. And two centuries later, Maxwell explained how a bolt of lightning is intimately related to the steady compass needle that has guided the path of ships for nearly a millennium.

We now know that a mere three distinct forces govern the natural world: gravity, the strong nuclear force, and the electroweak force. Some would remind us of the Higgs field and tell us that the correct number is four. Three is a manageable number, and even four isn't so much, but we scientists are an ambitious bunch. We hope to find a way to reduce that to two, or maybe even one. It seems that history tells us that if we try hard enough and dig deeply enough that we'll find more connections—ones that are hidden to us now. Perhaps, just perhaps, we will one day find that all of the known forces are different manifestations of a single, universal phenomenon.

Of course, by now you know all that. The question is "What comes next?" And there, the road ahead is murkier, but one does not have to have read very many popular science articles to think that the answer is imminent. Indeed, one often hears breathy reports that the answer is out there, made by an unheralded genius—currently unappreciated by the scientific community—who has had a crucial insight which will make everything clear. Claims are made that our current elegant and complex scientific edifice is about to neatly collapse into a single box and then will be presented to the multitudes wrapped with a single bow. The scientific establishment will have metaphorical egg on its collective face. Of course, this tale will probably not come to pass.

The idea of an unappreciated genius is deeply engrained in our culture, often one outside the academic community. For instance, we have been fed the myth of Einstein—a lowly patent clerk disconnected from the physics establishment—overturning the prevailing physics paradigm of his time in 1905. However, the truth is different. Einstein was a classically trained physicist. He completed his PhD dissertation at the end of April 1905, and it was accepted in July of that year. He had graduated with a bachelor's degree in math and physics in 1900, and he worked on his doctorate for the next five years. It is true that he worked in a Swiss patent office during some of those years—after all, he was married at the time, and he had a wife to support and bills to pay. But his academic training was world class. He most certainly wasn't an outsider.

1905 was indeed a year of spectacular intellectual achievement for Einstein. He wrote four seminal papers. The first was his paper on the photoelectric effect (received March 18, published June 9), which showed that light was not only a wave but also a particle. His second (received May 11, published July 18) explained "Brownian motion," which is the chaotic and jittery motion of dust particles that one sees in a microscope. The explanation is that the dust motes are pummeled by atoms, and thus this was the first direct evidence of the existence of atoms.

His third paper (received June 30, published September 26) laid out the principles of special relativity and made major changes to the laws of motion at speeds near the speed of light. And his fourth and final paper in 1905 (received September 27, published November 21) described the equivalence of mass and energy. It is this paper from which his famous equation $E = mc^2$ originates, although that equation is never cited in that paper in that form.

It was a decade later that Einstein caught his second wind and published his theory of general relativity—the one that showed that gravity could be described as the bending of space and time. Indeed, general relativity remains the most successful theory of gravity devised thus far and is the one taught to advanced physics students even today. It remains a remarkable achievement, even viewed from the vantage point of a century after the fact.

While it is inarguable that Einstein was a creative genius—indeed, a strong candidate in any contest to determine the most impactful physicist of all times—it is possible to overstate his impact on science. His paper on the photoelectric effect, which is the one for which he received his Nobel Prize in 1921, was one of many papers published that helped the physics community develop the field of quantum mechanics. It was an influential paper—one worthy of a Nobel Prize—but it did not lead the way, nor was it a singular contribution.

Einstein's contributions to his theories of relativity were more creative and unique, but even there he had help. The equations of special relativity are called the Lorentz equations, as they were independently developed in 1892 by George Fitzgerald and Hendrick Lorentz and were generally known in the physics community by 1905. And Henri Poincaré and Joseph Lamour noted before Einstein's first relativity paper was published that the equations implied that time would be experienced differently by different observers. The deep and insightful connection between Einstein's equations and the mixing of space and time was made clear in 1908 when Hermann Minkowski showed that relativity was most simply understood as describing a four-dimensional space, in which space and time were connected and interchangeable. The story is similar in 1915, when Einstein first published his theory of general relativity, which connected accelerations and gravity. The Minkowski formulation is what made it so clear that the bending of space and time is the real cause of what we call gravity.

Now none of this is meant to take away any of Einstein's seminal contributions. He was, without a doubt, a creative genius and he made substantial advances in our understanding of the laws of nature. But he was not an isolated genius. He lived in the thriving intellectual environment found in Europe at the beginning of the twentieth century. The idea that he was "just a patent clerk" is simply not true. Nor did he do it all, and he also spent the second half of his life following an unproductive path, without the success of his earlier years.

If we can dispense with the idea that a theory of everything is currently being fermented in the mind of a single individual, what about the idea that there exists a community out there that will, by virtue of their joint efforts, create the ultimate theory? Is that credible?

Well, it's certainly more likely than the idea that a single individual will solve everything. But here we need to look at the historical record to get a sense of the rate at which advances have been made. It rarely happens overnight and never out of context. For example, it was in the 1680s that Newton unified celestial and terrestrial gravity, and it was about two centuries later when Maxwell unified electricity and magnetism in 1865. The unification of electromagnetism and the weak nuclear force took about another century, with the key advances occurring in the late 1960s. (Of course, the weak force was only discovered in 1899, so perhaps it is unfair to say it took a century for the unification to occur.) And many of what are called "Maxwell's equations" were developed decades earlier by such other physics luminaries as Michael Faraday, Carl Friedrich Gauss, and André-Marie Ampère.

Thus, we see that, historically speaking, the time scale between individual unifying advances has been centuries, although it appears that the rate at which discoveries are being made might be speeding up. If we claim that it took only half a century for electroweak unification, then perhaps the grand unified theory might take half of that and perhaps a theory of everything might be half of that. If that were true, we're still talking about another forty years or so.

But that prediction is almost certainly wildly optimistic. As I hope I have persuaded you in earlier chapters, significant advancement in our understanding of the laws of nature will rely heavily on experiments that will illuminate connections that we do not yet appreciate. However, the scientific community faces a limitation, and that limitation is technology. We are limited in our ability to make advances toward a theory of everything by our slowing ability to measure the behavior of matter at higher and higher energies—at an ever-increasing set of temperatures. Let's take a look at that history.

The best way we know to generate very high energies and temperatures has been to use particle accelerators, and we have achieved a dizzying increase in our capabilities over the last century or so; indeed, the increase has been by more than a billion times.

The earliest particle accelerators were glass tubes from which much of the air was removed, leaving a respectable vacuum. Electric fields were applied to electrodes embedded in the tubes, with electrical potentials in the range of a few thousand volts up to eventually perhaps a hundred thousand volts. These first tubes were created in the mid-1800s. The discovery of X-rays in 1895 was achieved using a device called a Crookes tube, with a voltage somewhat above 5,000 volts. (The actual tube and voltage employed in that discovery are lost to history.)

It is perhaps valuable to remind you that the effective voltage of an accelerator is related to the energy of beam it can deliver. A 5,000 volt accelerator can deliver a beam of protons or electrons, with each particle carrying 5,000 electron volts of energy. The most powerful particle accelerator on the planet currently in operation is designed to create beams of protons with a staggering seven trillion electron volts of energy. However, the Planck energy, which we remember from Chapter 3 is the energy at which a theory of everything is thought to apply, is a much higher energy—a full 1.2×10^{28} electron volts of energy, about a quadrillion times more than the energies we can currently generate.

While the era of the Crookes tubes was the birth of particle accelerators, the twentieth century brought with it innovations in radio technology, which is exactly what was needed to make more powerful particle accelerators. Essentially, generating powerful oscillating electric fields is key. In addition, in 1930, Ernest Lawrence invented the cyclotron, which combined those same strong electric fields and magnets to make even more powerful particle accelerators. His first cyclotron was about five inches in diameter, and it accelerated particles to an impressive 80,000 electron volts.

Lawrence was an ambitious and gifted guy and he dreamed big. He improved his first cyclotron to the point that it broke the million electron volt barrier, but he wanted more. By 1939, he had built a cyclotron with a diameter of sixty inches, capable of generating energies of sixteen million electron volts. He was on a roll.

However, he encountered technical issues. The traditional cyclotron only works when the particle beams are moving slowly compared to the speed of light. As you may know, when objects move near the speed of light, they act as if their mass is increasing. This effect makes the cyclotron principle ineffective.

It is possible to correct for the effect of the increasing mass of particles by changing the frequency of the radio waves used to accelerate the beams of particles. Such a device is called a synchrocyclotron, and Lawrence built a huge one that began operations in 1946. It eventually achieved a beam energy of 350 million electron volts.

Synchrocyclotrons mitigated the problem of the rise of relativistic effects, but they suffered from other problems and a new technology was required.

The next step in accelerator technology is called a synchrotron, which is what is used in the Large Hadron Collider (LHC), the most powerful particle accelerator operating today. Synchrotrons use powerful magnets to make the beams of particles move in a circle. At a particular place on the circle, researchers place oscillating electric fields, which speed the beam particles up more and more as they pass through them. The magnets guide the beams around so that they pass again through the electric fields. Essentially the same electric field is used over and over again. The basic idea is shown in Figure 8.1.

Synchrotrons work best when the particle beams are moving at constant speed. Because particles approach the speed of light as they gain more and more energy, but never exceed it, their speed eventually becomes nearly constant, independent of energy. This allows accelerator scientists to use the same frequency oscillating electric fields. Because the beams travel at essentially a single speed, no matter what energy they have, they take the same amount of time to go around the ring and thus the orbits and the accelerating electric fields are synchronized.

However, there is an important effect that needs to be taken into account. As the beams become more energetic, it becomes harder and harder to make them move in a circular path. This isn't any different than what happens when you put a ball on a string and twirl it around your head, so it moves in a circle. The faster you twirl the ball, the more tension is on the string.

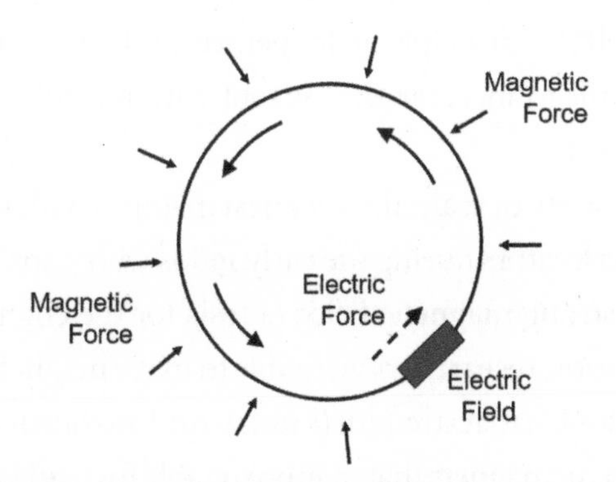

Figure 8.1 Synchrotrons use an array of magnets to make a beam move in a circular path so that they can repeatedly pass through an electric field that causes the beam to gain energy.

The same thing is true for synchrotrons. And, since it is the magnets that make the beam follow a circular path, as the energy of the beam increases, accelerator operators need to increase the strength of the magnets. Since the magnetic fields are made using electric currents, this means that they simply increase the current in the magnets. By this combination of synchronized beams and electric fields, and increasing magnet strength, researchers have been able to make very powerful accelerators.

However, it's worth mentioning that making strong magnets is very hard. To do so, you need very high electric currents, and to do that, you need to minimize the electric resistance in the wires in the magnets. If you cool the wires, this reduces the resistance and, if you reduce the temperature enough, the resistance of the wires becomes zero. This technique is what that has made modern accelerators possible.

You'd think that we'd be home free. It seems that we could simply increase the current higher and higher, making more and more powerful accelerators, but there is a magnetic field strength above which this principle no longer works. It becomes more and more difficult to increase the current with a resulting increased magnetic field.

To give a sense of scale, the very best magnets built for the use in particle accelerators during the early 1980s were capable of generating very strong magnetic fields (4 tesla for the cognoscenti). By the early 2000s, researchers were able to make magnetic fields that were only double that strength (8 tesla). And, in the early 2020s, the most powerful magnets that may be suitable for use in accelerators are only about double those of the 2000s (16 tesla). I say "may be suitable," because while prototypes exist, no accelerator using magnets of this strength has been built. We will return to this magnet problem in a bit.

There is one final technique that particle physics scientists can use to squeeze every bit of performance out of their equipment, and that's to take two beams traveling in the opposite direction and cause them to collide them head-on. Head-on collisions are far more powerful than ones in which you take a single beam and cause it to collide with a stationary target. The first colliding-beam accelerators were built in the mid-1960s, and they used beams of electrons. However, the most powerful accelerators today cause beams of protons to collide using large synchrotrons.

This book is not about the details of how accelerators are built, and I glossed over many details in my description of how they work. The real bottom line for those of us wanting to know about a theory of everything is that we want to know how long it will take us to build a particle accelerator powerful enough to generate Planck-scale energies.

The most obvious thing to do is to look at past performance and project out historical trends. Figure 8.2 shows the growth in energy of accelerators using protons over the last fifty years or so. The first thing you'll note is that the trend has changed over time. From the period of 1971 (ISR) to 1987 (Tevatron), the growth in energy was rapid, but recent years have brought a slower growth.

The LHC began operations in 2011, and it is designed to generate collisions at a maximum energy of 14 trillion electron volts (TeV). It is seven times more powerful than the earlier Tevatron. The next circular collider (NCC) is a proposed future accelerator that, if it is built, will begin operations no earlier than 2040 and far more likely it will be 2060. It is being designed to operate at 100 TeV, or about seven times more powerful than the LHC.

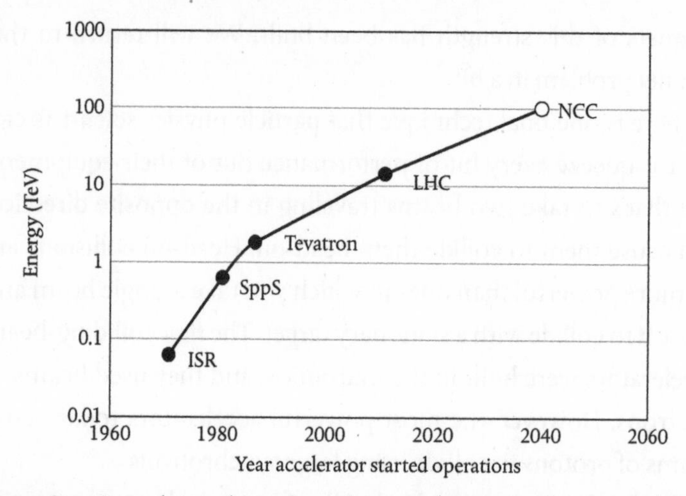

Figure 8.2 Growth in achieved accelerator energy as a function of year. The ISR (Intersecting Storage Rings accelerator) began operations in 1971, with a collision energy of 0.062 TeV, and the LHC (Large Hadron Collider) began operations in 2011 with a design collision energy of 14 TeV, an increase of a factor of a hundred. The NCC (next circular collider) is a proposed 100 TeV accelerator that does not yet exist and, if it is built, it will not operate until at least 2040 and possibly 2060. This figure only shows accelerators that involve protons or antimatter protons.

Let's take the optimistic 2040 date and see what recent trends in accelerator performance teach us. In rough numbers, it takes about thirty years to build an accelerator with a performance that is seven times higher than its predecessor. That means from the 2011 start date of the 14 TeV LHC, we can predict that the creation of a Planck-scale accelerator (1.2×10^{16} TeV) to begin operations in the year 2541. During that time, we will have built eighteen successive accelerators, each with an energy seven times more powerful than its successor. Sadly, unless my medical colleagues are about to announce some miraculous life-preserving advance, I will not be around to see it.

Of course, this simple calculation has ignored something very important—the size of the circular tunnel that houses the accelerator. The LHC is made of a ring with a diameter about 5 miles (~9 km) across. If the strength of magnets doesn't increase beyond those used in the LHC, the Planck-energy collider would have a diameter a quadrillion times wider—call it 10^{19} meters. That's a lot. The boundary between the solar system and interstellar space is the heliosphere and is a sphere about 2.5×10^{13} meters across. Thus, a Planck-scale accelerator would be 400,000 wider than that. It works out that the Planck-scale accelerator would have a diameter of about 1,000 light-years. This is, of course, absurd.

There is some slim hope. In those calculations, I have ignored the increase in strength of magnetic fields as technology develops. If we were able to continue doubling the strength of our magnets every thirty years, we wouldn't need quite as big an accelerator. Then it would only need to be about twice as wide as the heliosphere. Still big, but I suppose in the year 2541 we might have space travel worked out well enough to do that. Unfortunately, we know of no way to increase magnetic fields that much.

And, of course, there is the issue of amount of material and costs. The amount of material needed to build even this wildly optimistic accelerator, using recent trends in technology is simply staggering.

So all this really means is that making measurements at Planck-scale energies is not going to happen using a series of bigger and bigger circular accelerators. What other options do we have?

Well, it would be helpful if we had an accelerator that didn't have all of those expensive and hard-to-improve magnets. One idea is to go back to the basics and make an accelerator where the particles just move in a straight line. This is called a linear accelerator, and it is at some level a step backward. The Crookes tube of the late 1800s was a linear accelerator.

However, in the intervening century, scientists have developed a new acceleration technology. Instead of electric fields, they use plasma, which is kind of like a gas in which the individual atoms have had their electrons removed. The most familiar form of plasma is a neon light, but, of course, plasmas used for particle accelerators are different and certainly technically more complex. Plasma acceleration technology is relatively new and not yet used in any large particle accelerator, but proof of concept devices have shown to create accelerations about a hundred times more powerful than the commonly used radio-based accelerators.

Long term, scientists hope to achieve voltages of about 10 billion volts per meter. That would greatly reduce the size of accelerators achieving beam energies that are available today. However, to get to Planck energies, we're still talking particle accelerators that are 10^{18} meters, which is 100 light-years. So that's not really a viable option for probing the energy scale of a theory of everything, but it might help us push forward into the unknown, studying phenomena that are higher energy than we can currently study, but maybe only ten or a hundred times higher energy than is possible in the present day. That's certainly something.

By the way, I have described the limitations of current technology and you could be left with the impression that the situation is hopeless and that is absolutely *not* the impression I want to give. Indeed, I hope that I have convinced you in earlier chapters that it is only through a steady stream of experimental scientific inquiry leading to a series of discoveries that we will make progress. However, it is important to manage expectations. Furthermore, it points toward the very practical point that progress will require technological innovation. The key players in making progress toward a theory of everything are not just particle physicists like my colleagues and me. Equally important—perhaps more important—are those scientists developing new approaches to making accelerators and to others who devise new techniques to make ultra-precise measurements. And none of this would be possible without the contributions of engineers, programmers, and other technical professionals. To answer hard and important questions, it takes a village.

Final Thoughts

So where does this leave us? Where do we stand in the journey toward a theory of everything?

Well, we have a couple of theories that do a decent job explaining the behavior of familiar matter and energy. That's a huge success. We should not lose sight of how far we've come from the days when we cowered in caves, cringing as the god of lightning fought the demons that surrounded us. Our understanding of the cosmos is a thing of which all of humanity can be proud.

On the negative side, our best current guess is that the energy scale at which a theory of everything becomes obvious is a quadrillion times higher than what we can now explore. And a quadrillion is a lot, as our example of our distant ancestor we encountered in Chapter 3 taught us; the intrepid *Australopithecus afarensis* who had an expert understanding of her environment, but could never have imagined great white sharks, penguins, or the rings of Saturn. There is just too much distance between what we've achieved and where we want to go to predict with any reliability what the end of our journey will look like.

This tremendous gap effectively rules out any possibility that a theoretical model proposed in the present day has any real chance of being a success. After all, any surprising discovery, like warp drive, parallel dimensions, or any of a number of ideas invented in the fertile minds of science fiction writers, would completely overturn our current theoretical edifice. And even a more ordinary discovery—say an unanticipated new subatomic force, or a modification to the laws of motion or gravity turning out to be the explanation for the mysteries of dark matter—could change everything. It would surely change our current, neatly interconnected, theoretical framework, like the one seen in Figure 3.8.

No, if we hope to one day develop a successful theory of everything, we're going to have to do it the old-fashioned way—by studying the world around us, finding a shiny fact here and an intriguing clue there. Occasionally, we will notice a loose thread in our theories and find that a tug on it unravels the entire fabric of our understanding of the laws of nature, and we will then use that loose thread to reweave it into a newer and more beautiful tapestry, one that better represents the way the universe actually works.

I hope I've convinced you that the techniques of the past—those of slamming beams of subatomic particles together at higher and higher energies—are not a quick path forward. Before we can credibly explore the energies of the Planck scale, we're going to have to develop new technologies that give us more powerful tools. And there are plenty of historical examples that show us how this works, like when Galileo turned his telescope to the heavens and saw the moons of Jupiter. With that one observation, he uncovered the empirical evidence that overturned over a millennium of geocentric theory. And when Roentgen and Becquerel discovered X-rays and radioactivity, respectively, they were able to do so because of technical developments that preceded them.

So, if colliding beams are not a quick path forward to a final answer, should we forgo them entirely? Not at all. After all, we don't know at what energy some new phenomenon will manifest itself. It could be just two or three times higher than we can now explore, or it could be a million times higher. However, until we look, we won't know.

In this book, I have tried to focus on the practical—on the realities of real research. It's not fast. It's not easy. It relies on blood and sweat and tears. It relies on grit and determination and lots of hard work, and it benefits from moments of serendipity—moments when a surprising observation changes everything. And it requires patience, oh so much patience.

There is an important aphorism that applies here: "Life is not a destination, but a journey." While we should never forget our goal, we should not forget to be mindful of how we travel and what we see on the way. Finding a theory of everything is an outcome to which any fundamental physicist aspires to achieve, but it is in the exploration that we find joy.

If I might be permitted a bit of poetic metaphor, which is illustrated in Figure 8.3, scientists are explorers—mountain climbers of the peaks of knowledge. While we all aspire to explore the highest and most distant peak, we can't get there without crossing the foothills that come before it. Indeed, if the theory of everything is the final goal, we won't get there without cresting smaller—although still very challenging—peaks, like trying to find a theory of quantum gravity or finding if the strong nuclear force and the electroweak force have common origins.

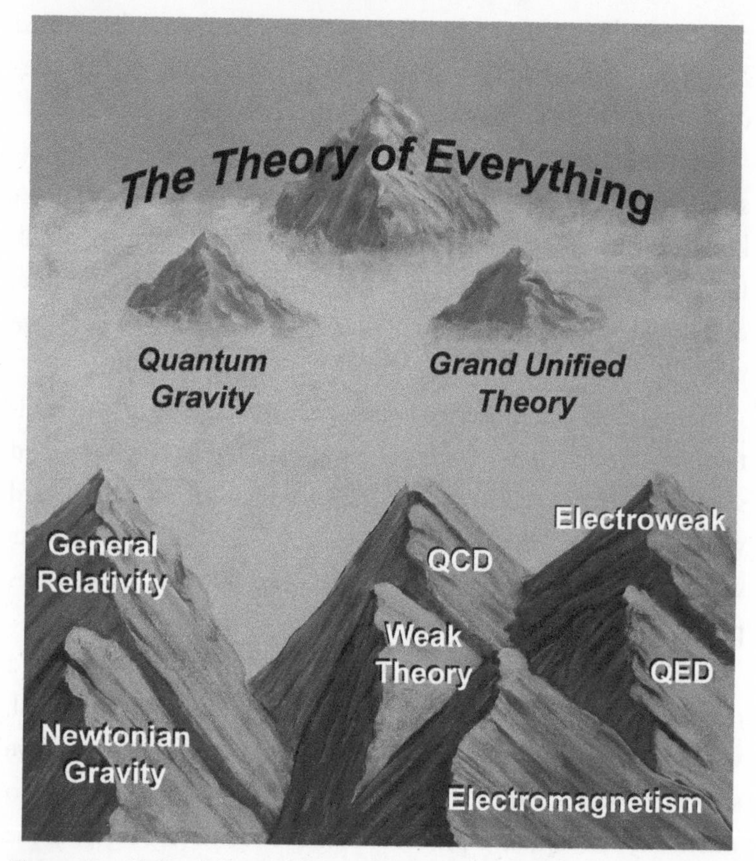

Figure 8.3 If the theory of everything is the pinnacle of knowledge, it is one that is far off in the distance, with the path to the summit hidden in the clouds of our ignorance and with well-explored foothills nearby. (Figure courtesy Diane Lincoln.)

But we're not there yet. Instead, what we've done thus far is to begin to explore the gentler knolls of Newtonian physics, relativity, and all of the hummocks of the standard model. The more distant peaks are shrouded by mists of our ignorance, waiting for intrepid explorers to dive in and move forward. No doubt many explorers will follow wrong paths and encounter crevasses too deep to cross, requiring them to turn back and find another way. And a lucky few will make discoveries that will help map the way for later generations of explorers to follow.

If you are a young reader of this book, and you're the kind of person who doesn't simply want—but literally needs—answers to the existential questions of science, it's a wonderful time to be alive and a fascinating scientific career lies before you, if you're so inclined. Over the last century we have discovered the quantum realm, radioactivity, and astonishing new ways to think of gravity. We can use powerful infrared telescopes to peer back to nearly the creation of the universe, to a time when the first stars flickered into existence. Using radio telescopes, we can peer even more deeply—to the moment when the fireball of the Big Bang cooled enough so light can pass unmolested across the cosmos.

In the subatomic realm, we can recreate temperatures and conditions that were last common a trillionth of a second after the Big Bang. We can literally melt protons and neutrons and study the nature of matter in a form that hasn't been observed since the beginning of time.

Even with all of those successes, you should not feel that there is nothing still to discover. We learned in Chapters 4 and 5 that the familiar matter of atoms makes up only a scant 5% of the matter and energy of the universe. Fully 95% is not understood. Even more humbling, we saw in Chapter 6 that something favored matter over antimatter by a tiny number. For every 2 billion antimatter particles, there were 2 billion and 1 matter particles. Without that very tiny asymmetry, we simply wouldn't exist.

That brings us to another important point. Our existence depends on many parameters of nature being just right for our existence. If gravity were stronger, the universe would have long ago collapsed in on itself. If gravity were weaker, it would have never collected the hydrogen of the universe into the stars that forged the elements that made us. And there are many more examples. Were the mass of the up and down quarks reversed, atoms would not exist. The cosmic coincidences go on and on.

In short, there are many unanswered questions about the nature of reality. Even more tantalizing, there are many unasked questions—questions that we cannot even fathom at this point.

Back in 2002, American Secretary of Defense Donald Rumsfeld was talking to reporters and said, "... as we know, there are known knowns; there are things we know we know. We also know there are known unknowns; that is to say we know there are some things we do not know. But there are also unknown unknowns—the ones we don't know we don't know..."

He was talking about military intelligence; however, his words apply equally well to exploration of the scientific unknown. While I've sketched out the known knowns and some of the more promising known unknowns, it's the unknown unknowns that will most likely reshape our understanding of the laws of nature.

I wish I could have written a book that told you what the theory of everything will be, but I can't. Nobody can. But I hope I have reminded you why it is interesting and have given you a concrete idea of how we'll move ahead—inch by inch, discovery by discovery—until one time, probably centuries, perhaps millennia, from now, when it will all fall into place. I won't see the final answer, and neither will you. That saddens me, but I take heart in knowing that I am part of a long, multigenerational effort to answer questions that were first asked before humanity learned how to write. And perhaps my colleagues and I will find some new clues and ask some questions that those who come after us will be able to pursue, and the cycle will begin yet again. That's why we get up every morning—to look again at the data and see what lessons our experiments and data are teaching us.

Now, if you'll excuse me, I have to get back to the lab. That last data plot had something weird in it. I wonder what it means . . .

EPILOGUE

As I wrap up this book, I have some parting thoughts.

While others may disagree, I do think that developing a theory of everything will one day be one of humanity's crowning achievements. However, I hope that now that you've finished reading this book, you have a realistic understanding of what the future might bring and the timescales that will be involved. Given the instrumentation available to us now, that ultimate theory—if it exists—is hidden from us, separated by a vast chasm of ignorance between where we are now and where we wish to go, and veiled by mists that only future technological advances will lift.

Unlike other books on the topic, I have emphasized that distance and the very real likelihood that future research will show us that all of our current guesses at what a theory of everything might look like are at best incomplete and perhaps downright wrong. For those who disagree with me and strongly believe that some current model, for example, superstring theory, is an accurate depiction of the final theory—what Einstein called "God's thoughts"—there is a Yiddish proverb that applies: Man plans, and God laughs.

I don't think that this position is a controversial one within the community of professional scientists who study these things, no matter what one reads in science popularizations about the theory of everything. I am very sorry to tell you that nobody can honestly tell you what the final answer will be.

Of course, this doesn't mean that general relativity or the standard model will be completely jettisoned; these two epic achievements have been too thoroughly tested and work too well. Any future theoretical improvement will make very similar predictions to our current favorites in the areas where we've tested them to date. But the answers to the questions of dark matter, dark energy, the matter/antimatter asymmetry, the puzzle of quark and lepton flavor, and other mysteries not covered in this book will undoubtedly change our understanding; of this, I am confident. Realistically, we can only wait and see what trail these discoveries will illuminate.

The bottom line is that future progress toward a theory of everything will proceed just as it has in the past, step by incremental step, slowly, but inexorably, leading to a fascinating future. It's a path I have followed and will continue to do so, knowing that I will not see the end. But I do hope that some young person will read this book and realize that these ancient questions are still interesting ones—ones worth investigating. And, in so doing, the torch will be passed to the next generation. If you are such a person, I bid you to go forward and do great things.

And, if you are a person who simply enjoys learning about the laws of nature, the future is a bright one. Over the years and decades, researchers will understand more and more, meaning that the rest of us can expect an unending series of discoveries to enjoy.

SUGGESTED READING

Chapter 1

Popular Science

E. Salaman. "A Talk with Einstein." *The Listener* 54 (1955): 370–371. The origin of Einstein's oft-quoted phrase "God's thoughts."

Chapter 2

Popular Science

P. Binetruy. *Gravity!: The Quest for Gravitational Waves.* Oxford: Oxford University Press, 2018. An introduction into gravitational waves and how they were discovered.

C. Calle. *Einstein for Dummies.* New York: For Dummies Press, 2005. A "For Dummies" book, written for the casual reader who wants to dig deeper.

R. Crease and C. Mann. *The Second Creation: Makers of the Revolution in Twentieth-Century Physics.* New Brunswick, NJ: Rutgers University Press, 1996. One of my favorite books on the history of particle physics. Well worth your time.

G. Gamow. *Mr. Tompkins in Paperback.* Cambridge: Cambridge University Press, 2012. A very accessible set of stories written by Gamow beginning in 1939, as well as later stories. There is no math, and the tales try to make relativity, quantum mechanics, and other topics accessible to a nonmathematical audience.

L. Lederman and D. Teresi. *The God Particle: If the Universe Is the Answer, What Is the Question?* Rev. ed. New York: Houghton Mifflin Harcourt, 2006. A lighthearted review of the history of particle physics, including several episodes in which Leon Lederman played a central role.

D. Lincoln. *Understanding the Universe: From Quarks to the Cosmos.* Rev. ed. Singapore: World Scientific, 2012. My rendition of the history of particle physics and cosmology.

D. Lincoln. *The Large Hadron Collider: The Extraordinary Story of the Higgs Boson and Other Things That Will Blow Your Mind.* Baltimore: Johns Hopkins University Press, 2014. The story of the Large Hadron Collider and how the Higgs boson was discovered.

I. Sample. *Massive: The Missing Particle That Sparked the Greatest Hunt in Science.* New York: Basic Books, 2013. The history of the development of the theory of the Higgs field and boson, along with the search and discovery. This is an excellent book.

G. Schilling. *Ripples in Spacetime: Einstein, Gravitational Waves, and the Future of Astronomy.* Cambridge, MA: Belknap Press, 2017. Written by a science popularizer, this book describes the search for and discovery of gravitational waves, which are a crucial confirmation of Einstein's theory of gravity.

D. Styer. *Relativity for the Inquiring Mind.* Baltimore: Johns Hopkins University Press, 2011. This is one of my favorite books to explain special relativity. It's a little challenging, but it's well written.

Technical

B. P. Abbott et al. "Observation of Gravitational Waves from a Binary Black Hole Merger." *Phys. Rev. Lett.* 116 (2016). The original paper in which gravitational waves were first observed.

C. W. Chou et al. "Optical Clocks and Relativity." *Science* 329 (Sept. 24, 2010): 1630. The original paper in which general relativity was confirmed by lifting a precise clock a single foot higher than a neighboring, identical clock.

F. Halzen and A. Marten. *Quarks and Leptons: An Introductory Course in Modern Particle Physics.* New York: Wiley, 1991. This is a graduate-level textbook that introduces the standard model. It is not for the fainthearted and requires a rather sophisticated physics and mathematical background.

C. Misner, K. Thorne, and J. Wheeler. *Gravitation.* Princeton, NJ: Princeton University Press, 2017. An advanced textbook on general relativity. Only for mathematically sophisticated readers.

D. Perkins. *Introduction to High Energy Physics*. 4th ed. Cambridge: Cambridge University Press, Cambridge, 2000. This is a college textbook that describes the standard model at an undergraduate level. It is well written and is ideal for the reader who has a decent grasp of calculus and at least a passing familiarity of differential equations. For the serious scholar, it's a great book.

Chapter 3

Popular Science

J. Baggott. *Quantum Space: Loop Quantum Gravity and the Search for the Structure of Space, Time, and the Universe*. New York: Oxford University Press, 2019. Book about loop quantum gravity by a science journalist.

S. Carroll. *Something Deeply Hidden: Quantum Worlds and the Emergence of Spacetime*. New York: Dutton, 2019. This book is about quantum mechanics, but it has some quantum gravity information. Carroll is a good writer.

M. Chalmers. "The Roots and Fruits of String Theory." *CERN Courier*, November 2018, p. 21. An interview with Gabriele Veneziano on the origins of superstring theory.

B. Greene. *The Elegant Universe: Superstrings, Hidden Dimensions, and the Quest for the Ultimate Theory*. 2nd ed. New York: W.W. Norton & Company, 2010. This book is a detailed description of superstring theory, aimed at a scientifically curious lay audience.

D. Hooper. *Nature's Blueprint: Supersymmetry and the Search for a Unified Theory of Matter and Force*. New York: Smithsonian Press, 2008. A book about supersymmetry. It's very well written, but slightly dated.

S. Hossenfelder. *Lost in Math: How Beauty Leads Physics Astray*. New York: Basic Books, 2020. This book is very critical of modern physicists and their bias that a future theory should reflect symmetries and be "beautiful." While there are certainly physicists like those she describes, it is not my experience that they are universal and not even all that common in the experimental community. Still, it's a book worth reading.

G. Kane. *Supersymmetry: Unveiling the Ultimate Laws of Nature*. New York: Basic Books, 2000. This book describes supersymmetry. It was written during a time when there was great optimism about the possibility that supersymmetry was right.

L. Lederman and C. Hill. *Symmetry and the Beautiful Universe.* New York: Prometheus Books, 2004. This book explains the role that symmetry has played in creating modern physics theories. The book is quite accessible.

See D. Lincoln. *Understanding the Universe* (reference for Chapter 2). This has an extensive discussion of how particle accelerators work.

See D. Lincoln. *The Large Hadron Collider* (reference for Chapter 2). This has an extensive discussion of how particle accelerators work.

C. Alden Mead. "Walking the Planck Length through History." *Physics Today* 54, no. 11 (2001): 15. This is Mead's letter to the editor responding to Frank Wilczek's commentary on the history of the importance of the Planck length.

C. Rovelli. *Reality Is Not What It Seems: The Journey to Quantum Gravity.* New York: Riverhead Books, 2018. A book about quantum gravity by one of the architects of the theory. Target audience is general public.

B. Schumm. *Deep Down Things: The Breathtaking Beauty of Particle Physics.* Baltimore: Johns Hopkins University Press, 2004. A book that talks about group theory in more detail than I have in this book, but still at an accessible level.

L. Smolin. *Three Roads to Quantum Gravity*. New York: Basic Books, 2001. Another popular science book about quantum gravity by one of the people who helped develop loop quantum gravity.

L. Smolin. *The Trouble with Physics: The Rise of String Theory, The Fall of a Science, and What Comes Next.* New York: Mariner Books, 2007. A book that is very critical of superstring theory. It's aimed at a general audience.

F. Wilczek. "Scaling Mount Planck I: A View from the Bottom." *Physics Today* 54, no. 6 (2001): 12. The first of three articles by a Nobel Prize–winning physicist on how the importance of the Planck length became evident over time.

F. Wilczek. "Scaling Mount Planck II: Base Camp." *Physics Today* 54, no. 11 (2001): 12. The second of three articles by a Nobel Prize–winning physicist on how the importance of the Planck length became evident over time.

F. Wilczek. "Scaling Mount Planck III: Is That All There Is?" *Physics Today* 55, no. 8 (2002): 10. The third of three articles by a Nobel Prize–winning physicist on how the importance of the Planck length became evident over time.

P. Woit. *Not Even Wrong: The Failure of String Theory and the Search for Unity in Physical Law*. New York: Basic Books, 2007. A book that is very critical of superstring theory. It gets its name from a quip by physicist Wolfgang Pauli. He was presented with a paper from a physicist, and he found it to be devoid of value. Sadly, he said, "It's not even wrong."

Technical

C. Alden Mead. "Possible Connection between Gravitation and Fundamental Length." *Phys. Rev.* 135 (1964): B849. This was Alden Mead's paper in which he showed that the Planck length is the smallest that current theories can describe. Note that this paper is for readers with a sophisticated understanding of physics.

J. Baez and J. Huerta. "The Algebra of Grand Unified Theories." *Bull. Am. Math. Soc.* 47 (2010): 483–552. This is a serious article, aimed at people with high-level math backgrounds, so it's not easygoing. However, if you are a serious math person, this will help you understand the interconnections between group theory and physics.

E. J. N. Wilson. *An Introduction to Particle Accelerators.* New York: Clarendon Press, 2001. This book is a textbook on how accelerators work. It is challenging, but accessible to a committed reader with some understanding of introductory physics.

A. Zee. *Group Theory in a Nutshell for Physicists.* Princeton, NJ: Princeton University Press, 2016. A book that discusses group theory in a mathematical context. It is a textbook, so it isn't light reading.

Chapter 4

Popular Science

J. Johnson Jr. *Zwicky: The Outcast Genius Who Unmasked the Universe.* Cambridge, MA: Harvard University Press, 2019. An in-depth biography of the pugnacious scientist who coined the term "dark matter."

D. Lincoln. "Dark Matter." *The Physics Teacher* 51(2013): 134. A very simple derivation of the rotation curves of an ideal galaxy. It can be easily followed by anyone with a single semester of algebra-based introductory physics.

J. Mitton and S. Mitton. *Vera Rubin: A Life.* Cambridge, MA: Belknap Press, 2021. A biography of Vera Rubin.

Technical

H. Andernach, trans. "English and Spanish Translation of Zwicky's (1933) The Redshift of Extragalactic Nebulae." arXiv:1711.01693v1 (2017). **A translation of Zwicky's 1933 paper in which he coined the phrase "dark matter."**

T. Clifton et al. "Modified Gravity and Cosmology." *Physics Reports* 513, no. 1–3 (March 2012): 1–189. **A comprehensive discussion of the many theories of alternative gravity. It's not for the faint of heart.**

M. Milgrom. "A Modification of the Newtonian Dynamics as a Possible Alternative to the Hidden Mass Hypothesis." *Astrophysical Journal* 270 (1983): 365–370. **Milgrom's first paper on modified Newtonian dynamics.**

CHAPTER 5

Popular Science

Note: there is considerable overlap in the popular science titles relevant for Chapters 4 and 5.B. Clegg. *Dark Matter & Dark Energy: The Hidden 95% of the Universe.* New York: Icon Books, 2019. **More information about dark matter and energy at a level similar to that in the book you are holding.**

R. Kirshner. *The Extravagant Universe: Exploding Stars, Dark Energy and the Accelerating Cosmos.* Princeton, NJ: Princeton University Press, 2014. **Similar material, but at a somewhat more technical level.**

R. Panek. *The 4% Universe: Dark Matter, Dark Energy, and the Race to Discover the Rest of Reality.* New York: Mariner Books, 2011. **Discussion of the search for dark energy and matter, with some extra emphasis on the history.**

Technical

O. Lahav, ed. *The Dark Energy Survey: The Story of a Cosmological Experiment.* Singapore: World Scientific, 2020. **A more technical treatment of one of the current leading dark energy experiments.**

S. Perlmutter. "Supernovae, Dark Energy and the Accelerating Universe." *Physics Today* 56, no. 4 (2003): 53. **Article written shortly after the discovery of dark energy by one of its discoverers.**

Chapter 6

Popular Science

G. Borissov. *The Story of Antimatter: Matter's Vanished Twin.* Singapore: World Scientific, 2018. Another antimatter primer, written by a colleague of mine, now a professor at University of Lancaster, UK.

F. Close. *Antimatter.* 2nd ed. Oxford: Oxford Press, 2018. A delightful tale of antimatter by an Oxford University emeritus professor. The book covers the material of this chapter, but in more detail.

M. Gardner. "The Fall of Parity." In *The Ambidextrous Universe*, 204–217. New York: New American Library, 1964. The tale of how the discovery of the nonconservation of parity in weak interactions unfolded.

B. Gato-Rivera. *Antimatter: What It Is and Why It's Important in Physics and Everyday Life.* New York: Springer, 2021. An antimatter primer written by a physicist at the Institute for Fundamental Research in Madrid, Spain.

L. Lederman and D. Teresi. *The God Particle: If the Universe Is the Answer, What Is the Question?* New York: Delta Books, 1993. Leon Lederman's book describes the experiment he did that killed parity conservation in pion decay, among many other things. There are newer versions than the one I cite here, but Leon autographed my copy when I was a young postdoc, so it has a special place for me.

G. L. Trigg. "Disproof of a Conservation Law." In *Landmark Experiments in Twentieth Century Physics*, 155–178. New York: Crane Russak, 1975. A series of essays describing important physics experiments of the twentieth century, including the Wu experiment.

Technical

J. Christenson et al. "Evidence for the 2π Decay of the K_2^0 Meson." *Phys. Rev. Lett.* 13 (1964): 138. The original paper by Cronin and Fitch that showed that CP was violated in neutral K meson decays.

T. D. Lee and C. N. Yang. "Question of Parity Conservation in Weak Interactions." *Phys. Rev.* 104 (1957): 254. Erratum: *Phys. Rev.* 106 (1957): 1371. The original paper by Lee and Yang where they realize that parity conservation in weak interactions had not been tested prior to January 1957.

A. Sakharov. "Violation of CP Invariance, C Asymmetry, and Baryon Asymmetry of the Universe." *Journal of Experimental and Theoretical Physics Letters* 5 (1967): 24–27. **Sakharov's paper on the three conditions for baryogenesis.**

C. S. Wu et al. "Experimental Test of Parity Conservation in Beta Decay." *Phys. Rev.* 105 (1957): 1413. **The original paper by Wu and all that killed parity conservation in beta decay.**

C. S. Wu and B. Maglich, eds. *Adventures in Experimental Physics. Gamma Volume* (pp. 101–123). Princeton, NJ: World Science Communications, 1973. **Chien-Shiung Wu's recollections of the fall of parity conservation in weak interactions.**

CHAPTER 7

Popular Science

B. Greene. *The Elegant Universe: Superstrings, Hidden Dimensions, and the Quest for the Ultimate Theory*. Repr. ed. New York: W.W. Norton, 2020. **The most famous of books describing superstring theory for a lay audience. It's very accessible.**

Technical

I. A. D'Souza and C. S. Kalman. *Preons: Models of Leptons, Quarks and Gauge Bosons as Composite Objects*. Singapore: World Scientific, 1992. **The only book on preons. It's moderately technical.**

H. Fritzsch and G. Mandelbaum. "Weak Interactions as Manifestations of the Substructure of Leptons and Quarks." *Physics Letters B* 102, no. 5 (1981): 319. **The original Fritzsch and Mandelbaum preon paper. (Quarks and leptons are made of two preons.)**

H. Harari. "A Schematic Model of Quarks and Leptons." *Physics Letters B* 86, no. 1 (1979): 83–86. **One of the easier preon models. (Quarks and leptons are made of three preons.)**

H. Harari and N. Seiberg. "The Rishon Model." *Nuclear Physics B* 204, no. 1 (1982): 141–167. **One of the easier preon models. (Quarks and leptons are made of three preons.)**

J. C. Pati and A. Salam. "Lepton Number as the Fourth 'Color.'" *Phys. Rev. D* 10 (1974): 275; Erratum: *Phys. Rev. D* 11 (1975): 703. This paper is the progenitor of both one form of preon and leptoquark theory.

M. A. Shupe. "A Composite Model of Leptons and Quarks." *Physics Letters B* 86, no. 1 (1979): 87–92. One of the easier preon models. (Quarks and leptons are made of three preons.)

CHAPTER 8

Popular Science

See D. Lincoln. *Understanding the Universe* (reference for Chapter 2). This has an extensive discussion of how particle accelerators work.

See D. Lincoln. *The Large Hadron Collider* (reference for Chapter 2). This has an extensive discussion of how particle accelerators work.

Technical

See E. J. N. Wilson reference for Chapter 3.

INDEX

For the benefit of digital users, indexed terms that span two pages (e.g., 52–53) may, on occasion, appear on only one of those pages.

Figures are indicated by *f* following the page number